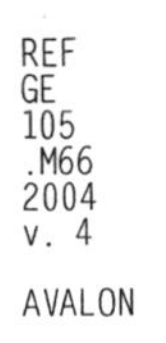

Teen Guides to

Environmental Science

Teen Guides to

Environmental Science

Human Impact on the Environment
Volume IV

John Mongillo
with assistance from Peter Mongillo

Greenwood Press
Westport, Connecticut • London

Library of Congress Cataloging-in-Publication Data

Mongillo, John F.
Teen guides to environmental science / John Mongillo with assistance from Peter Mongillo.
p. cm.
Includes bibliographical references and index.
Contents: v. 1. Earth systems and ecology—v. 2. Resources and energy—v. 3. People and their environments—v. 4. Human impact on the environment—v. 5. Creating a sustainable society.
ISBN 0–313–32183–3 (set : alk. paper)—ISBN 0–313–32184–1 (v. 1 : alk. paper)—
ISBN 0–313–32185–X (v. 2 : alk. paper)—ISBN 0–313–32186–8 (v. 3 : alk. paper)—
ISBN 0–313–32187–6 (v. 4 : alk. paper)—ISBN 0–313–32188–4 (v. 5 : alk. paper)
1. Environmental sciences. 2. Human ecology. 3. Nature–Effect of human beings on. I. Mongillo, Peter A. II. Title.
GE105.M66 2004
333.72—dc22 2004044869

British Library Cataloguing in Publication Data is available.

Library of Congress Catalog Card Number: 2004044869
ISBN: 0–313–32183–3 (set)
0–313–32184–1 (vol. I)
0–313–32185–X (vol. II)
0–313–32186–8 (vol. III)
0–313–32187–6 (vol. IV)
0–313–32188–4 (vol. V)

First published in 2004

Greenwood Press, 88 Post Road West, Westport, CT 06881
An imprint of Greenwood Publishing Group, Inc.
www.greenwood.com

Printed in the United States of America

The paper used in this book complies with the Permanent Paper Standard issued by the National Information Standards Organization (Z39.48–1984).

10 9 8 7 6 5 4 3 2 1

Contents

ACKNOWLEDGMENTS

The authors wish to acknowledge and express the contribution of the many nongovernment organizations, corporations, colleges, and government agencies that provided assistance to the authors in the research for this book. The authors are grateful to the Greenwood Publishing Group for permission to excerpt text and photos from *Encyclopedia of Environmental Science*, John Mongillo and Linda Zierdt-Warshaw, and *Environmental Activists*, John Mongillo and Bibi Booth. Both books are excellent references for researching environmental topics and gathering information about environmental activists. Many thanks to those who provided special assistance in reviewing particular topics and offering comments and suggestions: Sara Jones, middle school director for La Jolla Country Day School in San Diego, California; Emily White, teacher of geography and world cultures at the 5th grade level at La Jolla Country Day School, San Diego, California; Lucinda Kramer and John Guido, middle school social studies coordinators, North Haven, Connecticut; Daniel Lanier, environmental professional, and Susan Santone, executive director of Creative Change, Ypsilanti, Michigan.

A special thank you goes to the following people and organizations that provided technical expertise and/or resources for photos and data: Neil Dahlstrom, John Deere & Company; Francine Murphy-Brillon, Slater Mill Historic Site; Lake Worth Public Library, Florida; Pacific Gas & Electric; Energetch; Environmental Justice Resource Center; NASA Johnson Space Center; Seattle Audubon Society; John Onuska, INMETCO; Cathrine Sneed, Garden Project; Denis Hayes, president, Bullitt Foundation; Ocean Robbins, Youth for Environmental Sanity; Maria Perez and Nevada Dove, Friends of McKinley; Juana Beatriz Gutiérrez, cofounder and president of Madres del Este de Los Angeles—Santa Isabel; Mikhail Davis, director, Brower Fund, Earth Island Institute; Randall Hayes, president, Rainforest Action Network; Tom Repine, West Virginia Geologic Survey; Peter Wright and Nancy Trautmann, Cornell University; Mary N. Harrison, University of Florida; and Huanmin Lu, University of Texas, El Paso.

Other sources include Centers for Disease Control and Prevention, Department of Environmental Management, Rhode Island; ChryslerDaimler; Pattonville High School; National Oceanic and

Atmospheric Administration; Chuck Meyers, Office of Surface Mining; U.S. Department of Agriculture; U.S. Fish and Wildlife Service; U.S. Department of Energy; U.S. Environmental Protection Agency; U.S. National Park Service; National Renewable Energy Laboratory; Tower Tech, Inc.; Earthday 2000; Marilyn Nemzer, Geothermal Education Office; U.S. Agricultural Research Service; U.S. Geological Survey; Glacier National Park; Monsanto; CREST Organization; Shirley Briggs, Vortec Corporation; National Interagency Fire Center/Bureau of Land Management; Susan Snyder, Marine Spill Response Corporation; Lisa Bousquet, Roger Williams Park Zoo, Rhode Island; Netzin Gerald Steklis, International National Response Corporation; U.S. Department of the Interior/Bureau of Reclamation; Bluestone Energy Services; OSG Ship Management, Inc.; and Sweetwater Technology.

In addition, the authors wish to thank Hollis Burkhart and Janet Heffernan for their copyediting and proofreading support; Muriel Cawthorn, Hollis Burkhart, and Liz Kincaid for their assistance in photo research; and illustrators Christine Murphy, Susan Stone, and Kurt Van Dexter.

The responsibility of the accuracy of the terms is solely that of the authors. If errors are noticed, please address them to the authors so that corrections can be made in future revisions.

Introduction

Teen Guides to Environmental Science is a reference tool which introduces environmental science topics to middle and high school students. The five-volume series presents environmental, social, and economic topics to assist the reader in developing an understanding of how human activity has changed and continues to change the face of the world around us.

Events affecting the environment are reported daily in magazines, newspapers, periodicals, newsletters, radio, and television, and on Websites. Each day there are environmental reports about collapsing fish stocks, massive wastes of natural resources and energy, soil erosion, deteriorating rangelands, loss of forests, and air and water pollution. At times, the degradation of the environment has led to issues of poverty, malnutrition, disease, and social and economic inequalities throughout the world. Human demands on the natural environment are placing more and more pressure on Earth's ecosystems and its natural resources.

The challenge in this century will be to reverse the exploitation of Earth's resources and to improve social and economic systems. Meeting these goals will require the participation and commitment of businesses, government agencies, nongovernment organizations, and individuals. The major task will be to begin a long-term environmental strategy that will ensure a more sustainable society.

CREATING A SUSTAINABLE SOCIETY

Sustainable development is a strategy that meets the needs of the present without compromising the ability of future generations to meet their own needs. Many experts believe that for too long, social, economic, and environmental issues were addressed separately without regard to each other. In creating a sustainable society, there needs to be an integration of goals related to economic growth, environmental protection, and social equity. Some of these integrated sustainable goals include the following:

- Improve the quality of human life
- Conserve Earth's diversity

- Minimize the depletion of nonrenewable resources
- Keep within Earth's carrying capacity
- Enable communities to care for their own environments
- Integrate the environment, economy, and human health into decision making
- Promote caretakers of Earth.

OVERVIEW

Teen Guides to Environmental Science provides an excellent opportunity for students to study and focus on the integration of ecological, economical, and social goals in creating a sustainable society. Within the five-volume series, students can research topics from a long list of contemporary environmental issues ranging from alternative fuels and acid rain to wetlands and zoos. Strategies and solutions to solve environmental issues are presented, too. Such topics include soil conservation programs, alternative energy sources, international laws to preserve wildlife, recycling and source reduction in the production of goods, and legislation to reduce air and water pollution, just to name a few.

Major Highlights

- Assists students in developing an understanding of their global environment and how the human population and its technologies have affected Earth and its ecology.
- Provides an interdisciplinary perspective that includes ecology, geography, biology, human culture, geology, physics, chemistry, history, and economics.
- "Raises a student's awareness of a strategy called sustainable development that meets the needs of the present without compromising the ability of future generations to meet their own needs" (Bruntland Commission). The strategy includes a level of economic development that can be sustained in the future while protecting and conserving natural resources with minimum damage to the environment. People concerned about sustainable development suggest that meeting the needs of the future depends on how well we balance social, economic, and environmental objectives—or needs—when making decisions today.
- Presents current environment, social, and economic issues and solutions for preserving wildlife species, rebuilding fish stocks, designing strategies to control sprawl and traffic congestion, and developing hydrogen fuel cells as a future energy source.

- Challenges everyone to become more active in their home, community, and school in addressing environmental problems and discussing strategies to solve them.

ORGANIZATION

Teen Guides to Environmental Science is divided into five volumes.

Earth Systems and Ecology

Volume I begins the discussion of Earth as a system and focuses on ecology—the foundation of environmental science. The major chapters examine ecosystems, populations, communities, and biomes.

Resources and Energy

Currently, fossil fuels drive the economy in much of the world. In Volume II conventional fuels such as petroleum, coal, and natural gas are reported. Other chapters elaborate on nuclear energy, hydrogen energy, wind energy, geothermal energy, solar energy, and natural resources such as soil and minerals, forests, water resources, and wildlife preserves.

People and Their Environments

The history of civilizations, human ecology, and how early and modern societies have interacted with the environment is presented in Volume III. The major chapters highlight the Agricultural Revolution, the Industrial Revolution, global populations, and economic and social systems.

Human Impact on the Environment

Volume IV discusses the causes and the harmful effects of air and water pollution and sustainable solution strategies to control the problems. Other chapters examine the human impact on natural resources and wildlife and discuss efforts to preserve them.

Creating a Sustainable Society

Volume V focuses on the importance of living in a sustainable society in which generations after generations do not deplete the natural resources or produce excessive pollutants. The chapters present an overview of sustainability in producing products, preserving wildlife habitats, developing sustainable communities and transportation systems, and encouraging sustainable management practices in agriculture and commercial fishing. The last chapter in this volume considers the importance of individual activism in identifying and solving environmental problems in one's community.

PROGRAM RESEARCH

The five-volume series represents research from a variety of recurring and up-to-date sources, including newspapers, middle school and high school textbooks, trade books, television reports, professional journals, national and international government organizations, nonprofit organizations, private companies, businesses, and individual contacts.

CONTENT STANDARDS

The series provides a close alignment with the fundamental principles developed and reported in the President's Council on Sustainable Development and the learning outcomes for middle school education standards found in the North American Association for Environmental Education, the National Geography Standards, and the National Science Education Standards.

MAJOR ENVIRONMENTAL TOPICS

The *Teen Guides to Environmental Science* provide terms, topics, and subjects covered in most middle school and high schools environmental science courses. These major topics of environmental science include, but are not limited to:

- Agriculture, crop production, and pest control
- Atmosphere and air pollution
- Ecological economies
- Ecology and ecosystems
- Endangered and threatened wildlife species
- Energy and mineral resources
- Environmental laws, regulations, and ethics
- Oceans and wetlands
- Nonhazardous and hazardous wastes
- Water resources and pollution.

SPECIAL FEATURES

Tables, Figures, and Maps

Hundreds of photos, tables, maps, and figures are ideal visual learning strategies used to enhance the text and provide additional information to the reader.

Vocabulary

The vocabulary list at the end of each chapter provides a definition for a term used within the chapter with which a reader might be unfamiliar.

Marginal Topics

Each chapter contains marginal features which supplement and enrich the main topic covered in the chapter.

Activities

More than 100 suggested student research activities appear at the ends of the chapters in the books.

In-Text References

Many of the chapters have specially marked callouts within the text which refer the reader to other books in the series for additional information. For example, fossil fuels are discussed in Volume V; however, an in-text reference refers the reader to Volume II for more information about the topic.

Websites

A listing of Websites of government and nongovernment organizations is available at the end of each chapter allowing students to research topics on the Internet.

Bibliography

Book titles and articles relating to the subject area of each chapter are presented at the end of each chapter for additional research opportunities.

Appendixes

Four appendixes are included at the end of each volume:

- Environmental Timeline, 1620–2004. To understand the history of the environmental movement, each book provides a comprehensive timeline that presents a general overview of activists, important laws and regulations, special events, and other environmental highlights over a period of more than 400 years.
- Endangered List of U.S. Wildlife Species by State.
- Website addresses by classification.
- Government and nongovernment environmental organizations.

CHAPTER 1

Air Pollution: Outside and Indoors

Air pollution is a major health problem in the United States and throughout the world. An estimated 3 million people die each year from the effects of air pollutants. Medical researchers have linked high levels of air pollution to illnesses and diseases. These health problems include asthma, allergies, emphysema, chronic bronchitis, lung cancer, and heart attacks. Pollutants in the air cost citizens billions of dollars every year in health care and lost time at work.

In a major health study conducted by the American Lung Association, about 50 percent of the U.S. population is breathing unhealthy amounts of air pollution. The findings, released in a 2001 annual report, *State of the Air*, revealed that more than 142 million Americans live in places with high levels of ozone air pollution, commonly known as *smog*.

If you would like to find out about the air quality of your community, log on to www.stateoftheair.org.

COMPOSITION OF AIR

Air is the mixture of gases that forms Earth's atmosphere. Air supplies the gases needed for respiration and photosynthesis. Air also contributes to weather and climate conditions. Air contains hundreds of

TABLE 1-1 Bad Breathing and Good Breathing Cities

Some Cities on the Bad Breathing List	Some Cities on the Good Breathing List
Los Angeles–Riverside–Orange County, CA	Fargo, ND
Bakersfield, CA	Duluth, MN
Philadelphia, PA	Flagstaff, AZ
Atlantic City, NJ	Honolulu, HI
Atlanta, GA	Colorado Springs, CO
Houston–Galveston Brazoria, TX	Laredo, TX
Baltimore, MD	Lincoln, NE
Washington, DC	Des Moines, IA
Knoxville, TN	Salinas, CA
Wilmington, DE	Bellingham, WA

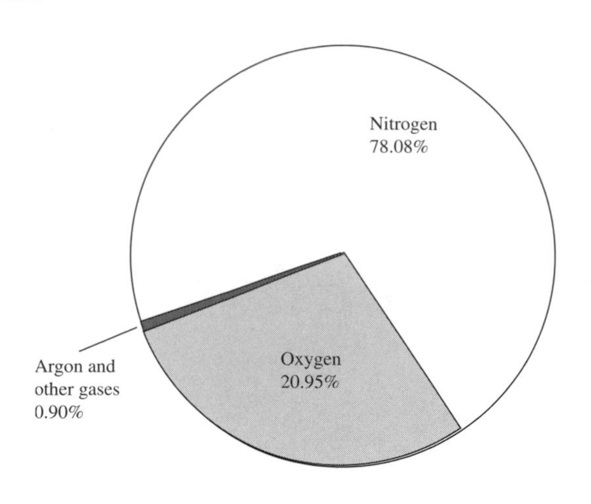

FIGURE 1-1 • The air surrounding the Earth contains hundreds of different gases. However, about 99.93 percent of air's mass is attributed to only three gases: nitrogen (N_2), which makes up about 78.08 percent, oxygen (O_2), which makes up about 20.95 percent, and argon (Ar), which makes up about 0.90 percent.

different gases. However, about 99.93 percent of air's mass is attributed to only three gases: nitrogen (N_2), which makes up about 78.08 percent; oxygen (O_2), which makes up about 20.95 percent; and argon (Ar), which makes up about 0.90 percent. The remaining percent of gases in air are called trace gases because they exist only in tiny amounts. Some of these include ozone (O_3), carbon dioxide (CO_2), water vapor (H_2O), and methane (CH_4). All these gases in the air are continually cycled through the environment by natural processes such as photosynthesis, *respiration*, and decomposition.

Refer to Volume I for more information about the atmosphere.

CAUSES OF AIR POLLUTION

Natural Causes

The composition of air contains natural pollutants that result from such events as forest fires, dust storms, and volcanic eruptions that release smoke and other pollutants into the air. Methane is produced by anaerobic bacterial decomposition. Other sources of air pollution include molds, spores, pollen, bacteria, and viruses. Even some plants such as coniferous trees release naturally occurring pollutants into the surrounding air. Oak trees give off a chemical, called isoprene, that can lead to the formation of smog and formaldehyde. However, most of these pollutants are not present in large enough amounts to cause serious pollution problems.

Human Causes

Human activities also contribute to air pollution. The major sources of air pollutants are caused by the burning of fossil fuels generated by transportation (trains, automobiles, trucks, and aircraft), power plants and factories, and other industrial activities. When fossil fuels go through the process of *combustion* in a motor vehicle's engine,

TABLE 1-2 Air Pollutants

Pollutant	Source
Carbon monoxide	Combustion of automobile engines
Fluorides	Manufacturing of fertilizer
Hydrogen sulfide	Oil wells, refineries
Lead	Leaded gasoline, lead smelting
Mercury	Pesticides, fungicides, chemicals
Nitrogen dioxide	Produced by combinations of nitrogen and oxygen during combustion of automobile engines, fertilizers
Particulate matter	Forest fires, fuel combustion, incineration, vehicles
Photochemicals	Photochemical reactions in the atmosphere
Sulfur dioxide	Combustion of fossil fuels, petroleum refining, smelting
Volatile organic compounds	Evaporation from solvents and liquid fuel
Greenhouse gases	Combustion of automobile engines
Ozone	Vehicles, factories, landfills

large amounts of particulates, carbon dioxide, sulfur dioxide, carbon monoxide, water vapor, and nitrogen dioxide are released into the atmosphere. Methane is caused by domesticated animals and from rice paddies. The amount of these pollutants that are emitted into the atmosphere is regulated and monitored by the Clean Air Act (CAA).

CLEAN AIR ACT

The CAA is a federal environmental law that regulates air emissions from stationary and mobile sources. The stationary sources include power plants and factories; the mobile sources include mobile vehicles, aircraft, and trains. Under this law, the U.S. Environmental Protection Agency (EPA) sets limits on how much of a pollutant can be contained in the air anywhere in the United States. This law ensures that all Americans have the same basic health and environmental protections.

The Six Criteria Air Pollutants

The CAA authorizes the EPA to establish *National Ambient Air Quality Standards (NAAQS)* to protect public health and the environment. As part of the NAAQS, the EPA has identified and established standards designed to protect human health and welfare for six pollutants known as the criteria pollutants. All these pollutants are common emissions from fossil fuel electric power plants, automobiles, and industries. There are six major pollutants:

- nitrogen oxides (NO_x)
- sulfur dioxide (SO_2)
- suspended particulates
- carbon monoxide (CO)
- lead (Pb)
- ozone (O_3) at ground level

NITROGEN OXIDES

Nitrogen oxides (NO_x) are gases containing nitrogen and oxygen that are emitted into the air primarily from the *emissions* of automobiles and power plants that burn petroleum and coal. Nitrogen oxides make up approximately 11 percent of all air pollutants that irritate the lungs, nose, and throat and also cause a variety of respiratory illnesses. One kind of nitrogen oxide is nitrogen dioxide (NO_2), which contributes to the formation of smog.

SULFUR DIOXIDE

Sulfur dioxide (SO_2) is a colorless gas with a characteristic acrid odor. It is a common pollutant emitted when fossil fuels containing sulfur are burned. Major emissions of sulfur dioxide in the United States are derived from power plants east of the Mississippi River, particularly those in the Ohio Valley. When released into the atmosphere, sulfur dioxide reacts with water vapor to form sulfuric acid. Sulfuric acid can be damaging to plants, aquatic ecosystems, and even structures made from rock and metal. Dry sulfate particles that are deposited on the ground react with moisture in soil to form sulfuric acid.

TABLE 1-3 Sources of Emissions from Nitrogen Oxides

Source	Percent
Motor vehicle exhausts	51%
Burning soft coal	46%
Nitric acid industries	1%
Solid waste disposal	1%
Miscellaneous	1%

The table illustrates the approximate amount of emission from nitrogen oxides from various sources as of 1992.

TABLE 1-4 Sources of Emissions from Sulfur Dioxides

Burning of high-sulfur coal	82%
Sulfuric acid industries	14%
Transportation	3%
Miscellaneous	1%

The table illustrates the approximate amount of emissions from sulfur dioxides.

High levels of sulfur dioxide gas cause significant constriction of air passages in the lungs, particularly among asthmatics, children, and elderly people. Exposure to high concentrations of sulfur dioxide pollutants can cause wheezing, shortness of breath, coughing, and respiratory diseases such as bronchitis.

PARTICULATE MATTER

Solid particles such as smoke, dust, ash, and soot that are present in the air are called *particulates*. Scientific evidence suggests that when inhaled, particulates can cause respiratory diseases and other health problems. Particulate matter in the air has been linked to between 10,000 and 100,000 premature deaths each year.

Most particulates in the air result from the burning of fossil fuels in coal-fired power plants and other industrial sources, and motor vehicles. Other sources of particulate matter include dust that is created by the clearing of land; smoke; soot resulting from forest or agricultural fires; and smoke, dust, and ash that result from natural events such as volcanic eruptions, windstorms, and tornadoes or other severe storms. Some particulates form in the air.

To remain suspended in air, particulates must be small and light in mass. Particle size is measured in micrometers, or *microns*, and may range between 0.005 to 100 microns. For comparison, the width of a human hair is about 100 microns. Particles smaller than 10 microns can travel deep into the respiratory system and become trapped on membranes within the lungs. The particles can cause excessive growth of lung tissue, leading to permanent injury. Because of the health problems associated with particulates, the size of the particles that can be released into the air by power plants, industrial processes, and automobiles is regulated by law. Any particle less than 2.5 microns in diameter, about 1/28th the width of a human hair, is considered a health hazard for humans by the EPA. Pollutants such as nitrogen oxides, sulfur dioxide, and particulates are measured in micrograms per cubic meter.

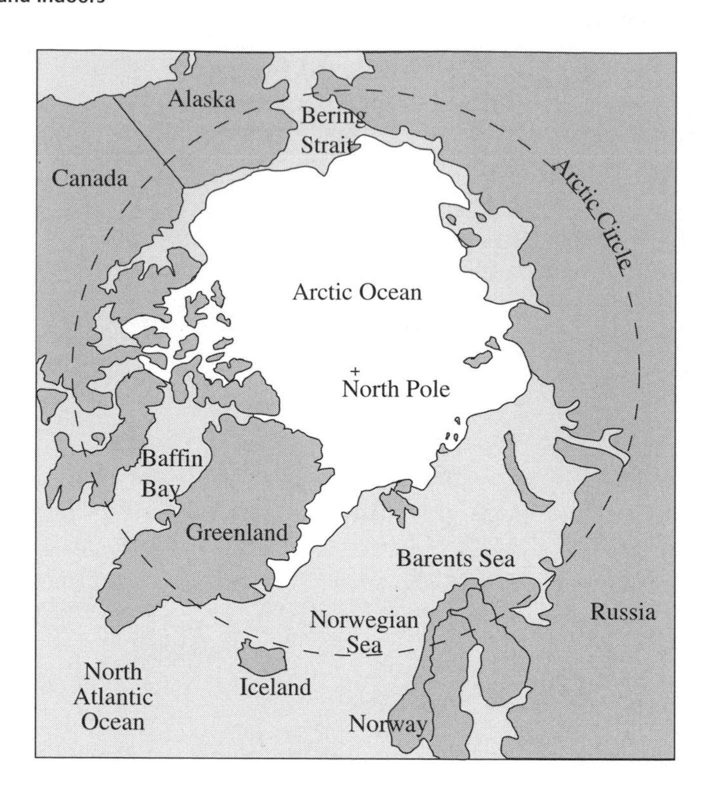

FIGURE 1-2 • The Arctic comprises the most northernmost portions of the Northern Hemisphere, including parts of Canada, Russia, the Scandinavian countries, and Greenland. The Arctic has experienced fallout of particulate matter such as dust and dirt from the atmosphere for centuries, leaving combustion products and chemical residues in the ice.

Air Pollution in the Arctic

A popular image of the Arctic depicts the region as remote, barren, and environmentally clean; however, this view of the Arctic does not represent modern conditions accurately. Despite the relatively small human populations of the Arctic (approximately 10 million inhabitants), the area has experienced pollution and other impacts from human activities since the Industrial Revolution.

The Arctic has experienced fallout of particulate matter such as dust and dirt from the atmosphere for centuries, leaving combustion products and chemical residues in the ice. The deposition of atmospheric particulates affects water, sediments, and soil, from which plants and animals take up these pollutants. In the 1970s researchers first observed an atmospheric feature known as Arctic haze. A haze that covers virtually the entire Arctic, this air pollution appears to originate largely from industrial areas of Europe and Asia, where coal- and petroleum-burning plants emit large quantities of particulates and haze-producing gases.

Carbon Monoxide

Carbon monoxide (CO) is different from carbon dioxide. Carbon dioxide is a gas that your lungs breathe out of your body. In contrast, carbon monoxide which is also a gas, is formed from the incomplete combustion of fossil fuels. The main source of carbon monoxide in the atmosphere comes from gasoline-powered motor vehicles.

Carbon monoxide gas is colorless, odorless, tasteless, and toxic. The gas is composed of one atom of oxygen and one atom of carbon. When present in the air, especially in an enclosed space, carbon monoxide is easily taken into the body, where it reacts with *hemoglobin* in the blood. In the blood, the carbon monoxide forms a compound (carboxyl-hemoglobin) that disrupts oxygen transport to the body's cells, robbing them of oxygen.

Symptoms of carbon monoxide poisoning include drowsiness, nausea, headaches, dizziness, and redness of the skin. High concentrations of carbon monoxide are deadly, particularly in children and people with heart disease or respiratory problems.

Sources of carbon monoxide in homes include unclean furnaces, fireplaces, and burning cigarettes. Use of carbon monoxide detectors can reduce the risk of carbon monoxide poisoning.

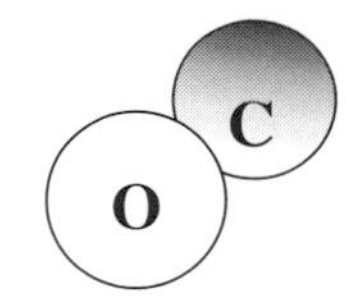

FIGURE 1-3 • A Carbon Monoxide Molecule High concentrations of carbon monoxide (CO) are deadly, particularly in children and people with heart disease or respiratory problems.

FIGURE 1-4 • A Carbon Dioxide Molecule Carbon dioxide (CO_2) exists in very small amounts in the atmosphere, making up only 0.04 percent of all atmospheric gases. Plants absorb carbon dioxide for use in photosynthesis; both plants and animals release it as a waste product.

OTHER AIR CONTAMINANTS

Some other air pollutants include lead, ground-level ozone, and volatile organic compounds. In the United States and in other developed countries, lead emissions in the atmosphere have dropped sharply with the phase out of leaded gasoline. However, lead emissions remain a major air pollutant in those countries using lead-additive gasoline. Ground-level ozone (found in the troposphere) is a major component of smog. Ozone is an oxygen molecule composed of three oxygen atoms. Ground-level ozone can be harmful to the human respiratory system. Ozone is released into the atmosphere when sunlight strikes the chemicals in the fumes that are emitted from automobile engines.

Refer to Chapter 3 for more information about ozone in the stratosphere.

Volatile organic compounds (VOC), compounds that easily evaporate, are potentially poisonous to humans and other organisms. Many VOCs are derived from petrochemicals, which are chemicals produced from petroleum. They include light alcohols, acetone, perchloroethylene, benzene, vinyl chloride, toluene, and methylene chloride. These volatile compounds are widely used in industry as solvents, degreasers, paints, thinners, and fuels.

When VOCs evaporate into the air, they are easily dispersed by wind, increasing the potential for exposure of humans to the substances. Excessive exposure to some kinds of VOCs can cause liver and kidney problems and cancer risks. These compounds also contribute to the formation of smog.

DID YOU KNOW?

Joggers who run outdoors when there is a high level of ozone pollutants can experience respiratory problems such as difficulty in breathing. Even children camping outdoors have been affected with respiratory problems as a result of high ozone pollution levels.

SMOG: A DIRTY HAZE IN THE ATMOSPHERE

Industrial Smog

Smog conditions can appear as a gray or brownish haze in the atmosphere. The grayish haze, also known as industrial smog, contains a mixture of particulate matter, sulfur, soot, and other contaminants. The smog

Refer to Chapter 5 for more information about toxic chemicals.

is usually located in highly populated and industrial areas where power plants and factories are centrally located. Smog can be a health hazard to humans, causing eye and lung irritation and possibly cancer. This kind of smog has been around since the Industrial Revolution began in the late 1700s when vast supplies of charcoal, and later coal and petroleum, were used as fuels.

Refer to Volume III for more information about the Industrial Revolution.

Photochemical Smog

Photochemical smog appears as a brownish haze and occurs in the air above urban centers that have heavy motor vehicle traffic. Photochemical smog originates when nitrogen oxides and *hydrocarbons* are emitted into the air by motor vehicles. Once the nitrogen oxides and the hydrocarbons are airborne, they undergo a photochemical reaction, accompanied by sunlight that produces nitrogen dioxide and ozone, a poisonous gas. The resulting smog becomes a health problem for humans, making breathing uncomfortable for those suffering from asthma and other respiratory diseases. Photochemical smog conditions have occurred in large cities such as Los Angeles and Mexico City.

Temperature Inversion

Both kinds of smog can be intensified by a condition known as temperature inversion. The smog continues to build up, causing harmful and dangerous air conditions at ground level. This occurs in the atmosphere when the air nearest to the ground is cooler than the air above it. The condition is a result of an increase in temperature high in the troposphere. Normally, surface air rises as long as the air around it is cooler. However, when the warm air is no longer less dense than the air above, it will stop rising. A temperature inversion can occur when a warm layer of air moves over the top of a cooler layer of smog. The warm air traps the cooler air so the smog cannot move away. The air becomes trapped with pollutants.

AIR POLLUTION AND HUMAN HEALTH

Once contaminants are in the air, exposure cannot easily be avoided. If high levels of outdoor air pollution are occurring in a city, for example, a large portion of the population will be exposed. Air pollution hurts the body both by directly inflaming and destroying lung tissue and by weakening the lungs' natural defenses. Exposure to air pollution can cause irritation of the eyes, nose, mouth, and throat, and it can cause coughing and sneezing. It can also worsen, causing lung diseases such as asthma, bronchitis, and emphysema. In some cases, it can even contribute to the premature death of people with heart and lung disease. For people who are already sick or especially sensitive, including young children, air pollution may mean discomfort, limited activities, increased use of medications, more frequent visits to doctors and

Lichens Gauge Air Pollution

A favorite food of caribou and reindeer, called *lichens*, is a very good indicator of pollutants in the air. Lichens absorb minerals, nutrients, water, and other substances from both solid and liquid substances as well as materials in the air. However, lichens cannot excrete any unnecessary substances, some of which are toxins, from the air, water, and soil. One such toxin is sulfur dioxide (SO_2), a major component of polluted air. The absorption of these substances in high amounts can cause the deterioration and the breakdown of the photosynthesizing unit of the lichens. Without photosynthesis, the lichens will die. Many lichens are too sensitive to exist in areas where sulfur dioxide is present in the air; or if they exist, they do so in small populations. Therefore lichens can be used as an indicator species to monitor and gauge air pollution levels in many areas.

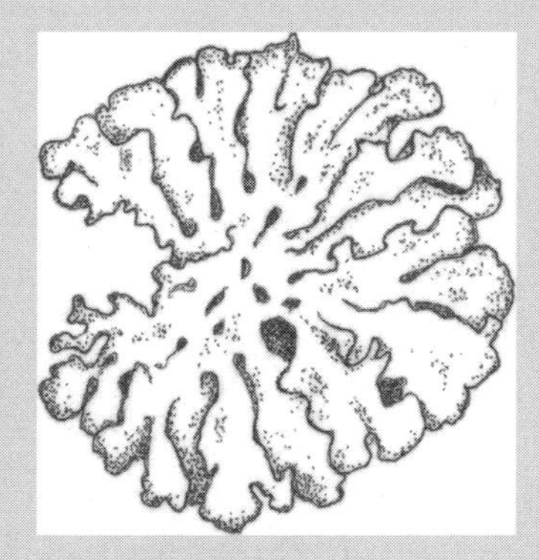

FIGURE 1-5 • There are different kinds of lichens. They can appear to be leafy, bushy, or crustlike. Lichens are referred to as an indicator species. When sulfur dioxide air pollution conditions exist, lichens cannot survive.

hospitals, and shortened lifespans. There is growing scientific evidence that suggests that air pollution has long-term effects on the lungs' ability to function and progresses the development of lung disease.

Over the past decade, there has been increasing recognition that minority and economically disadvantaged populations are disproportionately subjected to a variety of environmental health hazards, including air pollution. The nation's most severe air pollution problems, including those caused by ozone, are typically found in urban areas, which house many minority populations. Approximately 86 percent of African Americans and 90 percent of Hispanics live in urban settings, as compared to 70 percent of Caucasians. Researchers have found that higher percentages of African Americans and Hispanics than Caucasians also live in areas where there is a high level of particulate matter, carbon monoxide, ozone, sulfur dioxide, and lead.

Refer to Volume I, for more information about lichens.

TECHNOLOGY STRATEGIES TO CONTROL AIR POLLUTION

The CAA has established rules that industrial plants and automobile manufacturers must follow to reduce emissions of many hazardous air pollutants. As a result of the law, industries use a variety of technologies to control emissions. These technologies include catalytic converters, electrostatic precipitators, and scrubbers.

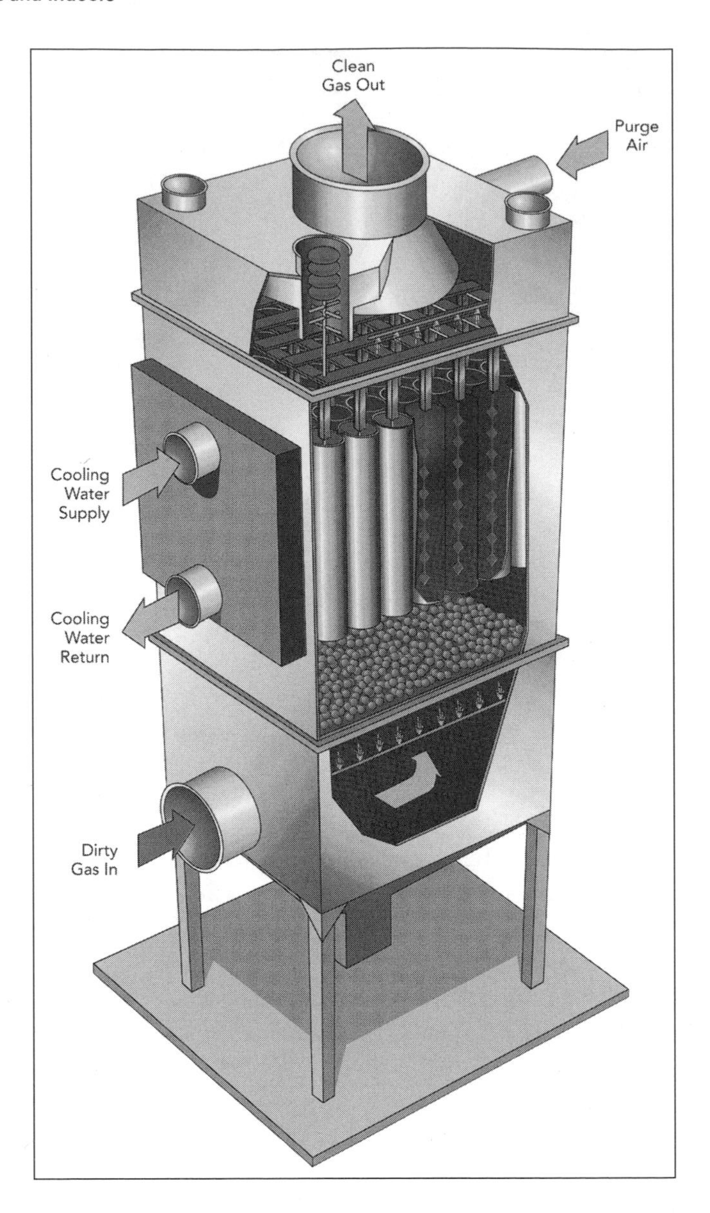

FIGURE 1-6 • The electrostatic precipitator uses electrostatic forces to remove particulates from gas streams before they are emitted into the atmosphere. The trapped particulates settle to the bottom where they are then removed. (Courtesy of Monsanto)

Catalytic Converters

The catalytic converter is equipment placed in the exhaust system of unleaded fuel, internal-combustion engines. The catalytic converter reduces the levels of air pollution from exhaust gases such as carbon monoxide. Catalytic converters are standard equipment in automobiles in the United States and many countries in Europe. As exhaust passes through the equipment, the metal *catalysts* in the converter oxidize carbon monoxide into carbon dioxide and water vapor. In many developing countries, where leaded gasoline is sold, cars are not equipped with catalytic converters. As a result, car exhausts produce much of the air pollution in those countries.

Electrostatic Precipitators

An electrostatic precipitator is a pollution control device that removes very small particles, known as particulates, from air emissions. In capturing particulates, these devices serve the same purpose as a dust filter, but they operate differently. Electrostatic precipitators use the physical principle of electrostatic attraction to capture particulates. Small particles can be given a positive or negative electrical charge by creating a sufficient electrical field. Once charged, particulates can be attracted to a collecting device, usually a plate or tube, of the opposite charge. Negatively charged particles will be attracted to a positively charged collector. This process of removing particulates is termed electrostatic precipitation. Electrostatic precipitators provide a very effective means of reducing particulates from emissions sources such as incinerators or industrial plants.

Scrubbers

A scrubber is a device that removes sulfur dioxide and other particulates from flue gases before they are emitted from tall stacks. The scrubbers are installed in coal-burning power plants, asphalt industries, concrete

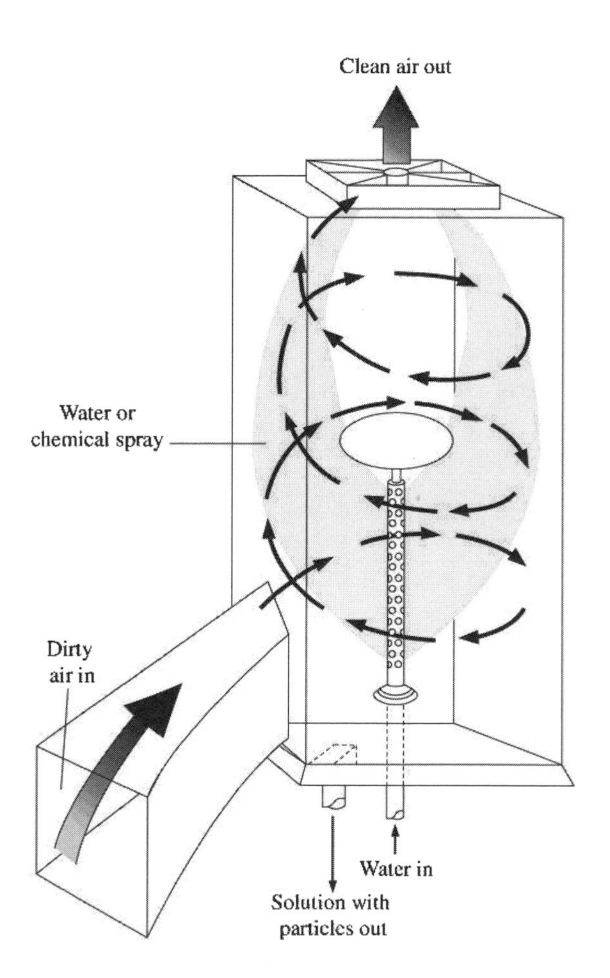

FIGURE 1-7 • Scrubber This illustration demonstrates how a typical scrubber is used to remove particulates from flue gases before they are emitted into the atmosphere.

factories, and a variety of other facilities that emit sulfur dioxides, hydrogen sulfides, and other gases. In the scrubber process, particulates, vapors, and gases are circulated and passed through a liquid solution containing lime. The lime reacts with the sulfur dioxide and forms sulfate compounds and other residue that are collected and disposed of in landfills.

INDOOR POLLUTANTS

Not all air pollution occurs outdoors. In fact, indoor air is often even more polluted than the air outside homes and workplaces. The EPA Science Advisory Board has ranked indoor air pollution among the top five environmental risks to public health. The U.S. Occupational Safety and Health Administration (OSHA) estimates that of the 70 million employees who work indoors in the United States, 21 million are exposed to poor-quality indoor air and millions more to environmental tobacco smoke. The pollution that occurs indoors may result from contaminants in faulty heating and air-conditioning units, gas stoves, cleaners and solvents, some kinds of carpeting, cigarette smoke, wall coverings, paints, and improperly stored chemical products.

Victims of indoor pollution may complain of one or more of the following symptoms: dry or burning nose, eyes, and throat; sneezing; stuffy or runny nose; fatigue or lethargy; headache; dizziness; nausea; irritability; and forgetfulness. Poor lighting, noise, vibrations, inadequate ventilation, pollutants such as VOCs, and psychological stress may also cause or contribute to these symptoms.

In attempting to cure *sick building syndrome* investigators or industrial hygienists examine four basic building factors that influence indoor air quality. The factors include the occupants; the heating, ventilation, and air conditioning system; possible pollutant pathways; and possible contaminant sources. The solution to the condition often lies in changing characteristics of a combination of these factors.

DID YOU KNOW?

Pot plants, such as chrysanthemums, spider plants, and English ivy, can remove certain indoor air pollutants.

TABLE 1-5 Some Major Indoor Air Pollutants

Pollutant	Source	Human Health Problems
Tobacco smoke—carbon monoxide, nitrogen dioxide	cigarettes	carbon monoxide—shortness of breath and heart failure nitrogen dioxide—emphysema and bronchitis
Radon	radon leaks through cracks in cellar walls	lung cancer
Formaldehyde	particleboard, plywood, and wood paneling	irritation of the eyes, nose, and throat; lung cancer
Asbestos	insulation in homes and schools	lung cancer, breathing difficulties

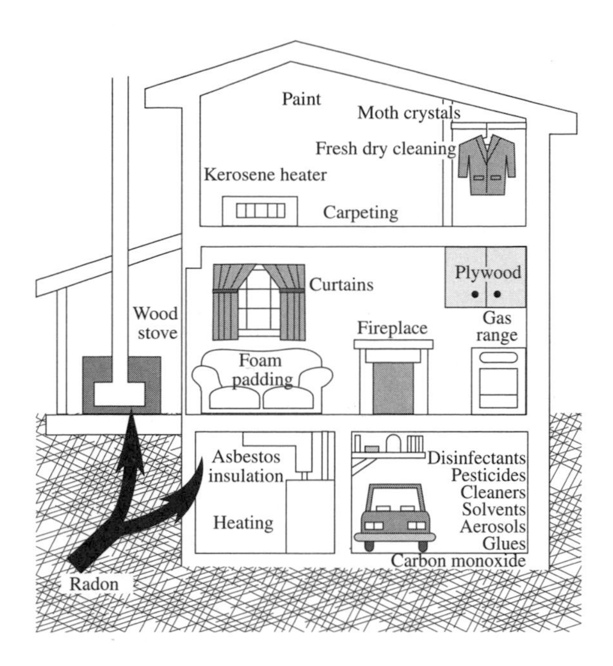

FIGURE 1-8 • Sick building syndrome is a condition in which building occupants experience acute health symptoms. Environmental illnesses may occur from poor lighting, noise, paints, disinfectants, furnishings, inadequate ventilation, or radon.

Asbestos

One kind of indoor pollutant is asbestos. Asbestos is a material that was widely used in the manufacture of a variety of products used in homes, schools, and industries. Asbestos is a very light, fibrous mineral that is easily carried in the air. These airborne asbestos particles can be inhaled by humans, causing health problems.

Based on studies linking asbestos to cancer and other diseases, the EPA banned the use of asbestos in insulation, fireproofing, or decorative materials in 1978. The next year, the EPA began a program to assist states in removing asbestos insulation that was flaking off pipes and ceilings in school buildings throughout the United States. This process was expanded in 1985, when the federal government passed the Asbestos Health Emergency Response Act (AHERA). This act required all elementary and secondary school buildings to be inspected to determine if materials containing asbestos had been used during their construction. If asbestos-containing material were found, the act required the materials to be contained or removed to prevent the asbestos fibers from breaking free and becoming airborne.

Radon

Another significant indoor health hazard results from radon (Rn) gas, which in many areas of the United States rises into buildings from underlying rock and groundwater. Radon is a radioactive element that exists as a colorless, odorless gas. The element forms naturally when *uranium* decays to radium, which further disintegrates to form radon gas. The byproducts of radon, one of which is polonium 218, can attach themselves to airborne dust and other particles. If inhaled, they can damage the lining of the lungs. In the outdoor environment, radon is not generally present in concentrations that pose a health risk. However, inside buildings with poor ventilation, radon can accumulate, creating a health hazard.

Radon may be one of the most dangerous of all indoor pollutants. The EPA estimates that radon gas is responsible for about 15,000 lung cancer deaths each year in the United States. After cigarette smoking, radon gas is the second leading cause of lung cancer.

In 1988, the EPA established a radon division that reported a wide distribution of radon gas in the basements of homes. The hazardous conditions resulted from radon gas seeping up through soil layers and entering buildings through cracks and openings in the foundations. If a building were sealed tightly and had poor ventilation, the gas would build up to harmful levels.

Homes and other buildings constructed on rocks and soil that contain granite and dark shales may have particularly high concentrations of radon gas. Such soils and rocks are common in the New England states and in New York, Pennsylvania, New Jersey, and the Midwest. The EPA advises homeowners in these regions to regularly have their homes tested for radon gas using a short-term (two to seven days') screening measuring device.

Radon Levels in Homes

The average outdoor radon level is about 0.4 picocuries per liter (pCi/L). The average indoor radon level is approximately about 1.3 pCi/L. If a home contains a radon level of 4 pCi/L or higher, the home should be tested for radon gas. Experts believe that homes with radon levels measuring a little less than 4 pCi/L should be tested as well.

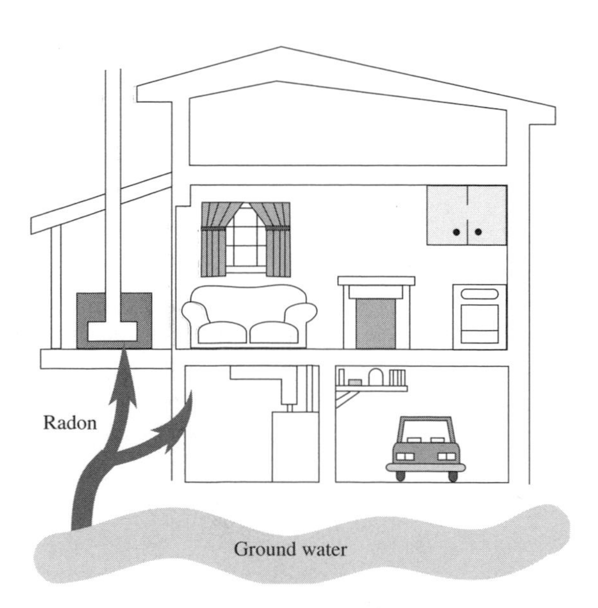

Figure 1-9 • Radon gas is one of the most dangerous indoor air pollutants. Radon can accumulate inside buildings that have poor ventilation systems causing health problems for occupants.

NOISE POLLUTION

Another kind of air pollution, noise pollution is any unwanted or unpleasant sound present in the environment. Sound and noise are measured in units of *decibels* (dBA). The noise level in an average house is about 40 dBA to 45 dBA; automobile traffic is between 55 dBA to 70 dBA. Exposure to sounds greater than 130 dBA for a length of time may cause hearing loss in humans. Exposure to excessive noise levels of greater than 80 dBA for a long period of time can cause both physiological and psychological effects on the body that result in anxiety, stress, and fatigue.

Reducing Noise Levels

In 1972, the U.S. Congress passed the Noise Control Act, which gave the EPA the authority to establish noise control regulations. Individuals or businesses found in violation of these regulations are subject to fines.

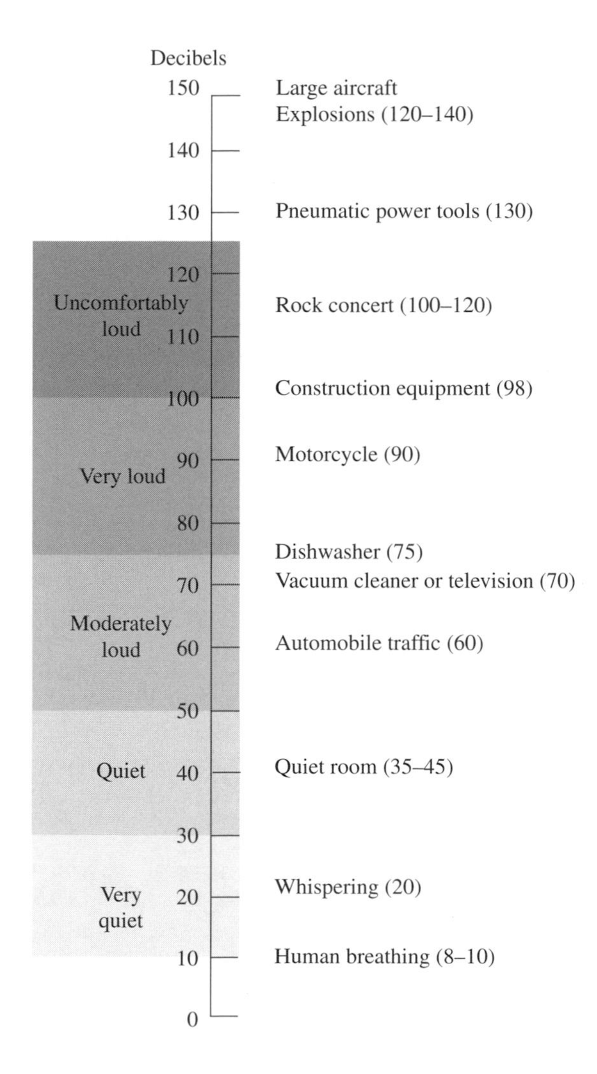

FIGURE 1-10 • Noise Pollution Sound and noise are measured in units of decibels (dBA). Exposure to excessive noise levels, of greater than 80 dBA for long periods of time, can cause both physiological and psychological damage on the body.

Since 1978, when Congress passed the Quiet Communities Act, more than 25 states have created nearly 1,000 noise-related ordinances.

The Department of Urban Development uses a guideline of 65 dBA as an acceptable level of day and night noise pollution. Some European countries are far ahead of the United States in developing quieter machines and tools and finding creative ways to reduce noise levels. However, with expanding populations and industrial growth, offensive noise levels are likely to continue to be a problem. The World Health Organization (WHO) reported that 100 million people are exposed to heavy traffic noise in excess of 65 dBA. This figure is above the accepted level of 55 dBA recommended by the WHO.

Vocabulary

Catalyst A substance such as vanadium that is used to speed up a chemical reaction but does not change its own form or composition in the reaction. Enzymes are biological catalysts.

Combustion A rapid chemical burning of a substance with oxygen to produce energy.

Decibel A numerical scale of the relative loudness of sound.

Emissions Pollutants released into the atmosphere.

Hemoglobin The protein in red blood cells that carries oxygen in the lungs to the body tissues.

Hydrocarbon A chemical compound containing hydrogen and carbon.

Lichen A plant that is formed by the symbiotic relationship between algae and fungi.

Microns A unit of length equal to one-millionth of one meter. The diameter of airborne particles are measured in microns (e.g., PM_{10} are particulate matter of 10 microns).

National Ambient Air Quality Standards (NAAQS) Measurable levels of pollutants permissible in the outside air throughout the United States as established by the EPA.

Particulates Dust and smoke particles that affect air quality.

Respiration The act of breathing for animals and humans.

Sick Building Syndrome A health condition where many people working in a building feel ill.

Smog A foglike condition containing contaminants and pollutants from power plants, internal combustion engines in motor vehicles, and domestic fires that can be hazardous to human health and other living things.

Uranium A rare heavy metal that can be manufactured as a fission fuel for nuclear reactors.

Activities for Students

1. Choose five cities from the Bad Breathing List and investigate what has caused them to be on the list by visiting the www.stateoftheair.org and EPA websites.
2. Visit or contact your local power plant and find out how they meet the requirements of the CAA.
3. Go to an industrial supply store or contact OSHA to find out what gear is available to protect professionals and laypeople from the air contaminants and VOCs that are found at the job site and in the air.
4. Ask a science teacher for the instruments that measure sound. Measure the decibels (dBA) throughout the day at school and at home. What causes these sounds? What causes sound decibels at a level of 130 dBA or greater?
5. Acquire a copy of *A Citizen's Guide to Radon* mentioned in *Books and Other Reading Materials* and design an activity to test your environment for radon levels.

Books and Other Reading Materials

Benarde, Melvin A. *Asbestos: The Hazardous Fiber.* Cleveland: CRC Press, 1990.

EPA booklet, *A Citizen's Guide to Radon: What It Is and What to Do about It* (OPA-86-004). Washington, DC. Contains general information on radon and how to reduce concentrations of radon gas in homes.

Glassman, Michael. *Pollution of the Environment: Can We Survive?* New York: Globe, 1974.

Graedel, Thomas E., and Paul Crutzen, Contributor. *Atmosphere, Climate and Change.* Scientific American Library Paperback, No. 55. New York: W. H. Freeman & Co., 1997.

Saign, Geoffrey C. *Green Essentials: What You Need to Know about the Environment.* San Francisco: Mercury House, 1994.

Somerville, Richard C.J. *The Forgiving Air: Understanding Environmental Change.* Berkeley: University of California Press, 1998.

Websites

American Meterological Society, *History of the Clean Air Act*, http://www.ametsoc.org/AMS/sloan/cleanair/index.html

Cars and Their Enviromental Impact, http://www.environment.volvocars.com/ch1-1.htm

Environmental Protection Agency, *The Plain English Guide to the Clean Air Act*, http://www.epa.gov/oar/oaqps/ peg_caa/pegcaain.html

Environmental Protection Agency Indoor Air Quality Homepage, http://www.epa.gov/iaq

Environmental Protection Agency Office of Air and Radiation, http://www.epa.gov/oar State and Territorial Air Pollution Program Administrators, Association of Local Air Pollution Control Officials, http://www.4cleanair.org

National Institute for Occupational Health (NIOSH), http://www.cdc.gov/niosh/homepage.html

Noise Pollution Clearinghouse, http://www.nonoise.org

OAR, *National Air Quality and Emissions Trends Report*, http://www.epa.gov/docs/oar/oarhome.html

Radon in Earth, Air, and Water, http://www.epa.gov/iaq/radon

U.S. Environmental Protection Agency Fact Sheet (EPA 400-F-92-004, August 1994), "Air Toxics from Motor Vehicles," http://www.epa.gov/oms/02-toxic.htm

U.S. Global Change Research Program, Carbon Cycle Science Program, http://www.geochange.er.usgs.gov/pub/carbon/

Agencies and Organizations

Toxic Substances Control Act Assistance, Information Service TSCA Hotline, Environmental Protection Agency, 401 M Street, SW, Mail Code 1230-C, Washington, DC 20460. (202) 554–1404. Free publications available from the hotline include *Asbestos Fact Book*.

CHAPTER 2

Global Air Pollution: Acid Precipitation, Climate Warming, and Ozone Depletion

Scientists study the structure, temperature, and composition of Earth's atmosphere to understand the dynamics of global air pollution. From their research, they have learned

- which pollutants cause the most problems
- how the pollutants are distributed throughout the atmosphere
- how long the pollutants remain in the atmosphere
- why Earth's atmosphere is going through changes, and how these changes have affected our environment, weather, and climate

ATMOSPHERE

The atmosphere is a protective layer of gases that surrounds Earth and plays a key role in Earth's ecosystem, *biogeochemical cycles*, and climate. As you learned in Chapter 1, the atmosphere contains the gases nitrogen (78.1 percent) and oxygen (20.1 percent) and other trace gases that include water vapor, argon, carbon dioxide, neon, helium, methane, hydrogen, ammonia, carbon monoxide, and ozone. The atmosphere also contains particulate matter, criteria pollutants, and aerosols.

Refer to Volume I for more information about Earth's atmosphere and biogeochemical cycles.

Earth's Troposphere

Earth's atmosphere is divided into several layers. However, most environmentalists are concerned mainly with two layers—the troposphere and the stratosphere. The troposphere is a region of the atmosphere that is closest to Earth. It includes water vapor and clouds and is a source of Earth's weather and *climate*. The troposphere is also the place where most of Earth's air pollution is found.

The troposphere extends from the surface to an altitude of about 18 kilometers (11 miles), although this height varies with latitude. Temperatures decrease with altitude in the troposphere. As warm air rises, it cools while falling back to Earth—a process, known as convection. Huge air movements mix the gases in the troposphere very efficiently. The chemicals, particulate matter, and other substances in the troposphere can be washed back to Earth by precipitation in the form of rain, snow, sleet, or fog.

Earth's Stratosphere

The stratosphere is a region that extends from 18 kilometers (11 miles) to about 50 kilometers (30 miles) above Earth. Commercial airlines fly in the lower stratosphere, where there are strong steady winds and little water vapor.

The temperatures in the stratosphere increase with altitude and are warm because of an *ozone* layer, a kind of shield that absorbs ultraviolet radiation from the sun. Nearly 90 percent of Earth's ozone is in the stratosphere. However, scientists are concerned that certain chemicals such as chloroflurocarbons (CFCs) that rise into the stratosphere are depleting the ozone layer. The outer layers of the atmosphere include the mesosphere, thermosphere, and exosphere.

Refer to Chapter 1 for more information about ozone in the troposphere.

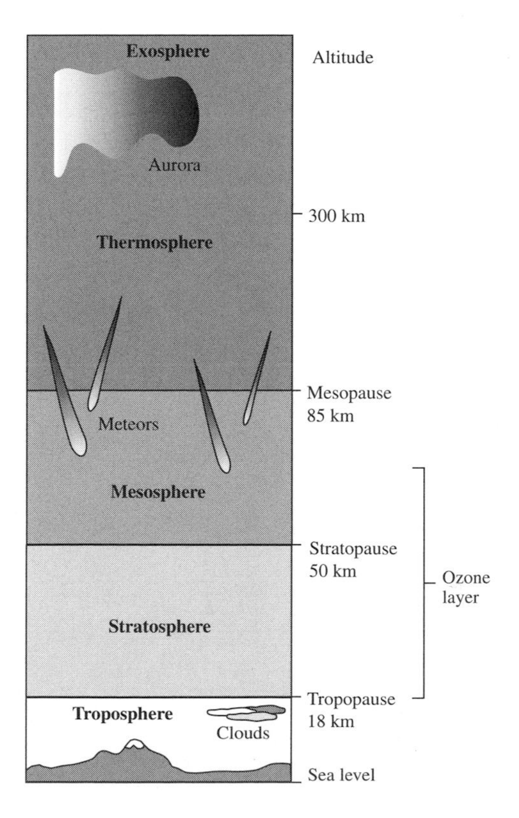

FIGURE 2-1 • Atmosphere The atmosphere is the protective layer of gases that surrounds the Earth and plays a key role in Earth's ecosystem and biogeochemical cycles. The ozone layer in the atmosphere is most important, because it provides a measure of protection to plant and animal life on the Earth's surface because it absorbs ultraviolet radiation.

ACID RAIN

In the troposphere, air pollution is a major cause of acid rain. Acid rain is precipitation that contains a higher level of acid than normal rainfall. Although usually called acid rain, this kind of precipitation occurs as snow, fog, or dry, airborne acidic particles that fall to Earth.

Acid rain is caused from particles released into the atmosphere by natural processes such as forest fires and volcanic eruptions. Acid rain also is derived from emissions resulting from the combustion of fossil fuels in automobiles and in boilers and furnaces in power plants and factories. When fossil fuels burn, they discharge particles such as nitrogen oxides (NO_x) and sulfur dioxide (SO_2).

Automobiles and other gasoline-powered vehicles emit a great deal of nitrous oxide (NO_2) through their exhaust systems. These nitrogen emissions dissolve in the water vapor of the atmosphere to produce nitric acid (HNO_3), which falls to Earth as acid precipitation.

Sulfur emissions are the primary source of acid rain. Electric utility power plants burn petroleum and coal, which release sulfur dioxide particles into the atmosphere. The particles mix with water vapor in the air to form sulfuric acid that eventually falls to Earth. Electric power plants account for about 70 percent of annual SO_2 emissions and 30 percent of NO_x emissions in the United States. Approximately 65 percent of the acid in all precipitation is made up of sulfuric acid. The remaining percentage is made up of nitric acids and other acids.

Composition of Acid Rain

What is an acid? It requires an explanation to understand how and why acid rain has an adverse impact on the environment. Scientifically, an acid is a chemical that releases hydrogen ions (H^+) into a solution.

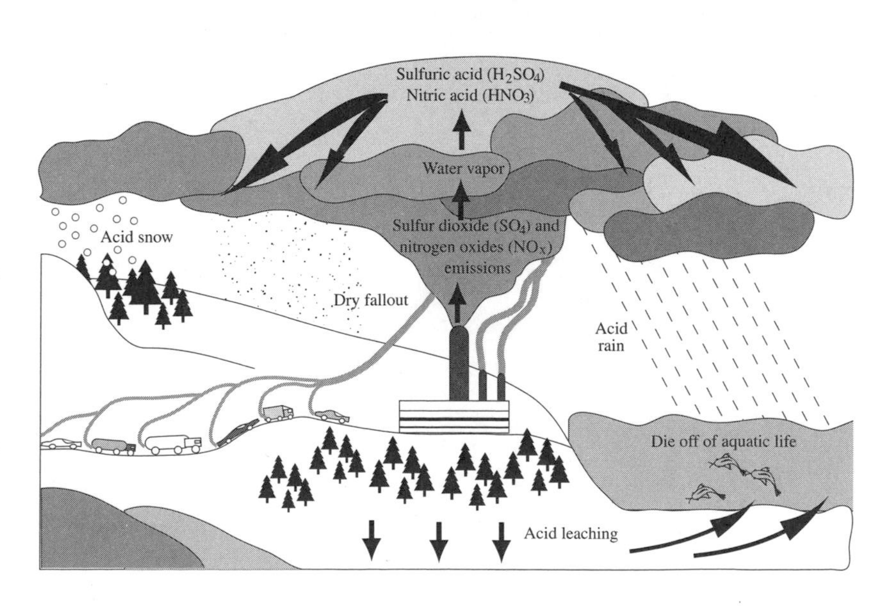

FIGURE 2-2 • Acid Rain Acid rain is caused by natural processes and human activities. Acid rain derives from emissions resulting from the combustion of fossil fuels in automobiles, homes, industries, and electric utility power plants. When fossil fuels burn, they discharge pollutants such as nitrogen oxides (NO_x) and sulfur dioxide (SO_2).

Some common properties of acids are that they have a sour taste, turn blue litmus paper red, and are soluble in water. In an acidic solution, the larger the number of hydrogen ions, the stronger the acid. Strong acids include hydrochloric acid (HCl), nitric acid (HNO_3), and sulfuric acid (H_2SO_4). Hydrochloric acid is used to clean metals. A diluted form of hydrochloric acid is found in the gastric juices of the human stomach. Nitric acid is used in making fertilizers and gun powder, and sulfuric acid is used in car batteries and for making fertilizers.

Measuring Acid Rain

Scientists measure the acidity of water using a pH reading tool. The pH scale ranges from 0 to 14, where 0 is the most acidic, 7 is neutral, and 14 is the most alkaline. A solution with a pH of 5 is 10 times more acidic than a solution with a pH of 6. Acid rain would have a pH of less than 5.6. Knowing the pH readings of water and soil is important. For example, trout or salmon cannot survive in water with a pH value of 4.0. Normal rainfall generally has a pH of about 5.6 because it contains a weak acid—carbonic acid (H_2CO_3).

According to the U.S. Environmental Protection Agency (EPA), the eastern United States now has rainfall with an average pH of 4.5, which is harmful to aquatic life and forests and is damaging to statues and buildings. In 1978, rainfall having a pH of 2.0—5,000 times more acidic than normal—was recorded in Wheeling, West Virginia.

DID YOU KNOW?

Most cola sodas have a pH reading between 3.7 and 4.0. They contain carbonic acid, which is a weak acid. Other weak acids include vinegar and lemon juice.

Global Impact of Acid Rain

Acid rain resulting from industrial activities is a significant problem in much of the northeastern United States, eastern Canada, Europe, Germany, Scotland, and Scandinavia. Other countries suffering from some acid rain conditions include eastern Brazil, Venezuela, southern India, and parts of China and Japan. The prospect of increasing consumption of coal in Asia worsens the global acid rain problem. For example, India's energy requirements to fuel much of its economy may triple its current level of coal consumption by 2020. The results will

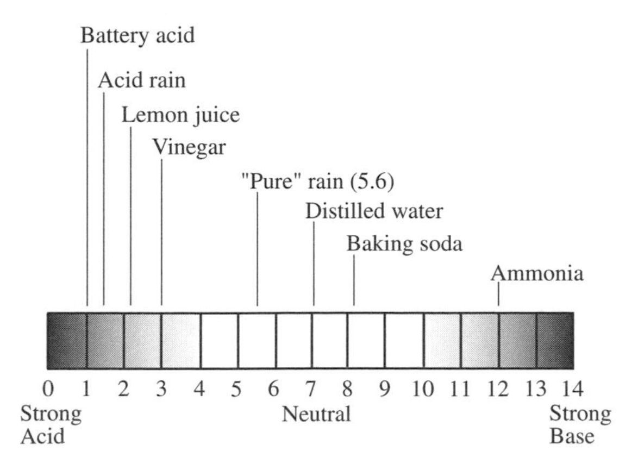

FIGURE 2-3 • The pH Levels of Various Substances

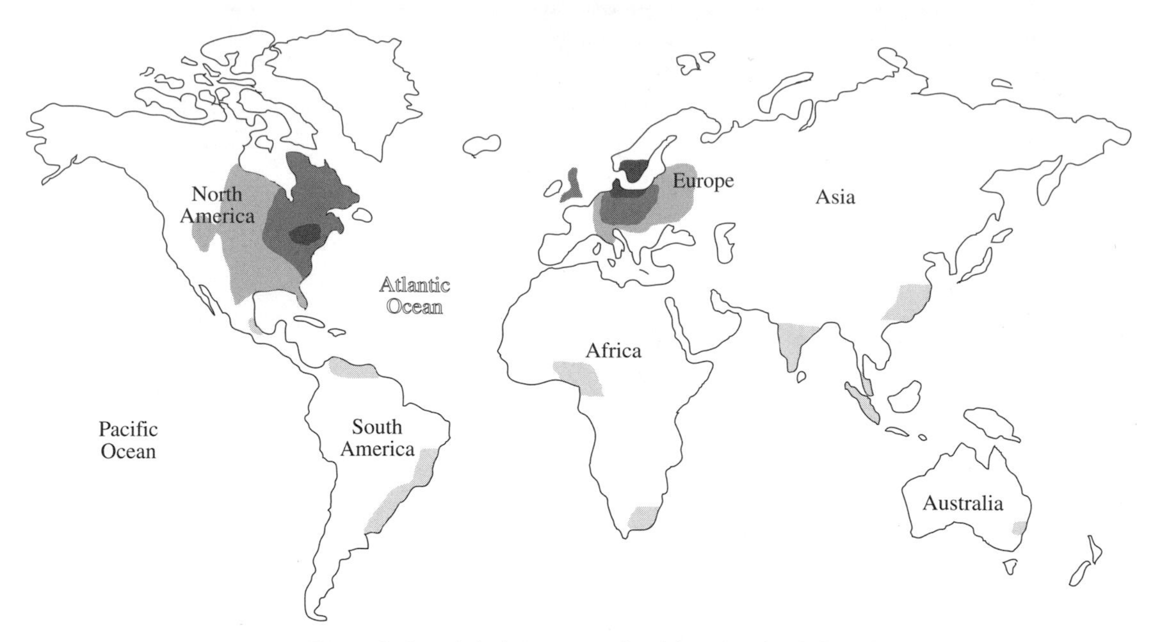

FIGURE 2-4 • Global Pattern of Acid Rain The darkest shaded areas on the map indicate the most severe acid rain conditions. The lighter shaded areas range from light to moderate acid rain conditions.

produce heavy acid rain conditions over many highly sensitive areas in the northeast region and coastal areas of India.

Other industrialized areas of Asia, Indonesia, Malaysia, the Philippines, and Thailand will also experience heavy acid rain deposition by 2020. However, China will be the most vulnerable of all the Asian countries because China depends heavily on coal as an energy source. Also, Chinese coal contains very high levels of sulfur, a major contributor to acid rain. By 2020, most of China's eastern locations will be suffering from acid rain problems unless conditions improve.

IMPACT OF ACID RAIN ON FORESTS

Acid rain is corrosive. Therefore, long-term exposure to liquid and dry acid particles can damage and destroy forests. Acid rain levels have caused premature deaths of many tree species in the United States, Canada, Norway, Sweden, southwest Poland, northwest Czech Republic, and southeast Germany. In sections of the Black Forest of Germany, conifers such as pine, fir, and spruce as well as some deciduous trees such as beech, maple, and oak are leafless. The trees are dead, damaged, or deformed from acidic conditions.

IMPACT OF ACID RAIN ON SOILS AND ECOSYSTEMS

When acid rain falls on soils containing alkaline substances, such as particles of limestone and sandstone, the acid is neutralized. In contrast, when acid rain falls on soils composed of igneous or metamorphic rocks, which do not contain the minerals or alkalis that can neutralize

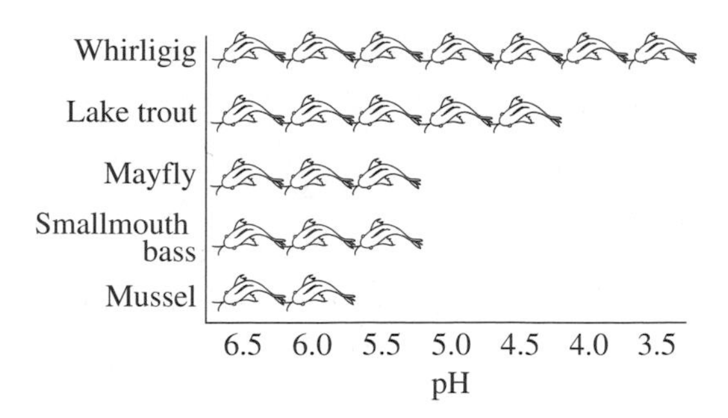

FIGURE 2-5 • Impact of Acid Rain on Aquatic Life

the acids, the acidity of the soil increases. Acids in soil can undergo chemical changes that are harmful to plants. As a result of these conditions, the plants may be damaged or destroyed. Wildlife is also harmed by acid rain. Animals feeding on these plants are then harmed as well. Acid rain also can *leach* vital nutrients, such as potassium and calcium, from the soil, making it less fertile.

Acidic buildup in lakes disrupts the reproductive processes of fishes and amphibians. Acids can also disrupt food chains. Studies have shown that lakes having a pH between 6.0 and 6.5 can support a variety of organisms; however, when the pH drops to 4.7 or less, microorganisms begin to die, with large-scale fish kills soon following. Hundreds of lakes in Canada and the northeastern United States (especially upstate New York) are acidified to the point where they contain little or no life. Many lakes in Scandinavia and central Europe are in the same condition.

In addition to destroying ecosystems and wildlife habitats, acid rain also damages human-made structures. Acid causes particles in statues, monuments, and buildings to chemically weather, particularly if the structures contain limestone or marble—rocks easily dissolved by acids. Such activity is evident in the Parthenon in Greece, which has weathered to the point of major decay through the corrosive nature of acids.

Reducing Acid-forming Emissions

Presently several methods are used to reduce acid rain damage caused by sulfur dioxide emissions. As you learned in Chapter 1, scrubbers and electrostatic precipitators are placed inside industrial smokestacks to remove many harmful emissions before they are released into the atmosphere. Using low-sulfur fuels and adding limestone to fuels also reduce sulfur dioxide emissions.

Spreading crushed limestone or lime, both alkalis (high pH values), over acidic lakes has helped to neutralize acidic water. But this method is very expensive and requires many repeated treatments.

Fluidized Bed Combustion

One technology used in reducing sulfur dioxide emission in coal-burning combustion is called fluidized bed combustion. Fluidized bed combustion is used for burning low-grade coal in a boiler that traps sulfur dioxide before being emitted into the atmosphere. The technology

Liming is used to treat acidic water in a lake. (Courtesy of Sweetwater Technology, Teemark Corporation)

was created through research and development sponsored by the U.S. Department of Energy.

In the fluidized bed combustion process, excess air is blown in from underneath the boiler. A mixture of pulverized coal and limestone is forced into the boiler, where it "floats" on the air while it burns. The calcium and some magnesium from the limestone mixture absorb the sulfur dioxide from the sulfur materials in the coal. As a result, the coal burns cleaner using this process.

No Borders for Acid Rain

Controlling acid rain distribution is complicated. Even with the technologies to reduce acid rain emissions locally, wind-swept pollutants can travel hundreds or thousands of kilometers from their sources. Pollutants released in one state or country are often deposited as acid rain in another. Acid rain problems in Canada and the northeastern United States, for example, have been traced to emissions from industries in the midwestern United States. Back in 1981, the Quebec Ministry of Environment informed U.S. officials that about 60 percent of the sulfur dioxide polluting Canada's air and water was coming from the United States. In Europe, emissions from industries in England and Germany fall as acid rain in the Scandinavian countries.

Controlling acid rain requires local and international cooperation. In 1989, the United States and 27 other nations signed a UN agreement that requires a reduction in nitrogen oxide emissions. In the United States, amendments to the Clean Air Act (CAA) included provisions for significantly reducing industrial emissions of sulfur dioxide and nitrogen oxides.

Refer to Chapter 1 for more information about the Clean Air Act.

SOME GOOD NEWS ON ACID RAIN REDUCTION

In 2002, there were some signs of progress in the reduction of acid rain emissions. Europe and the United States have enforced laws to reduce emissions of sulfur dioxide and nitrogen dioxides. Using catalytic converters, burning cleaner coal, and generating electricity by using natural gas have curbed the release of pollutants. Through these methods, the United Kingdom has reduced acid rain emissions by 50 percent in the last 15 years.

GREENHOUSE EFFECT AND GLOBAL WARMING

Greenhouse Effect

As you learned earlier, Earth is surrounded by a thin atmosphere consisting of gases such as carbon dioxide, dust particles, and water vapor. The gases in the atmosphere act like glass in a greenhouse: Solar energy passes through the atmosphere and is absorbed at Earth's surface. The radiant heat that reflects back into the atmosphere is then trapped by these gases, which prevent it from escaping too quickly back into space. This greenhouse effect is what makes Earth warm and habitable, and it is vital to life. Also, the energy resulting from the greenhouse effect is essential to the heating of the ground surface, the melting of ice

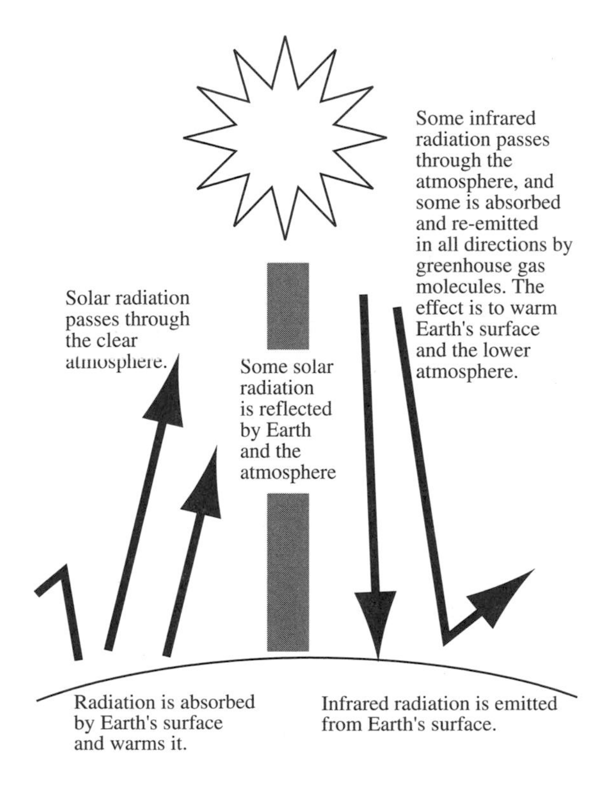

FIGURE 2-6 • Without the greenhouse effect, the Earth's average global temperature would be −18°C (−0.4°F), rather than its current 15°C (59°F).

and snow, and the evaporation of water. Without the greenhouse effect, Earth's average global temperature would be −18°C (−0.4°F), rather than its current 15°C (59°F). Even the oceans would freeze under such conditions.

Greenhouse Gases

The greenhouse effect is made up of specific atmospheric gases that trap heat. These greenhouse gases are produced from natural sources such as volcanic activities, forest fires, and changes in the photosynthesis and respiration rates in various ecosystems. However, greenhouse gases are also produced from human activities such as the burning of fossil fuels. Scientists acknowledge that the buildup of these gases in the atmosphere is trapping too much of Earth's heat.

As a result of the greenhouse gases building up in the atmosphere, Earth is growing warmer. The gradual rise of temperature over the entire Earth's surface caused by the greenhouse effect is called global warming. Five major greenhouse gases impact global warming:

- carbon dioxide
- nitrogen oxides (NO_x)
- chlorofluorocarbons (CFCs)
- methane
- water vapor

Carbon Dioxide

Carbon dioxide (CO_2) is by far the most important greenhouse gas. Carbon dioxide, along with nitrogen oxide, is emitted into the atmosphere every time fossil fuels, such as petroleum and gasoline, are burned for energy by automobiles, power plants, and factories. Increases of carbon dioxide also occur when large tracts of forests are cut down and burned.

Approximately half of the carbon dioxide released into the atmosphere from the burning of fossil fuels is captured by forests during photosynthesis or absorbed by oceans, which serve as an enormous carbon sink or reservoir. The other 50 percent of the carbon dioxide escapes into the atmosphere and remains there for many decades. The result is that over time, the amount of carbon dioxide increases and builds up in the atmosphere. Before the Industrial Revolution, about 220 years ago, the average concentration of carbon dioxide in the atmosphere was about 275 *parts per million (ppm)*. In 1958, carbon dioxide increased to 315 ppm; and by 1996, the amount reached 361 ppm. By 2001, carbon dioxide rose to 366.7 ppm in the atmosphere. When atmospheric carbon dioxide levels rise, the global temperatures also rise. It is estimated that 76 percent of the increase in global temperature warming is caused by carbon dioxide alone.

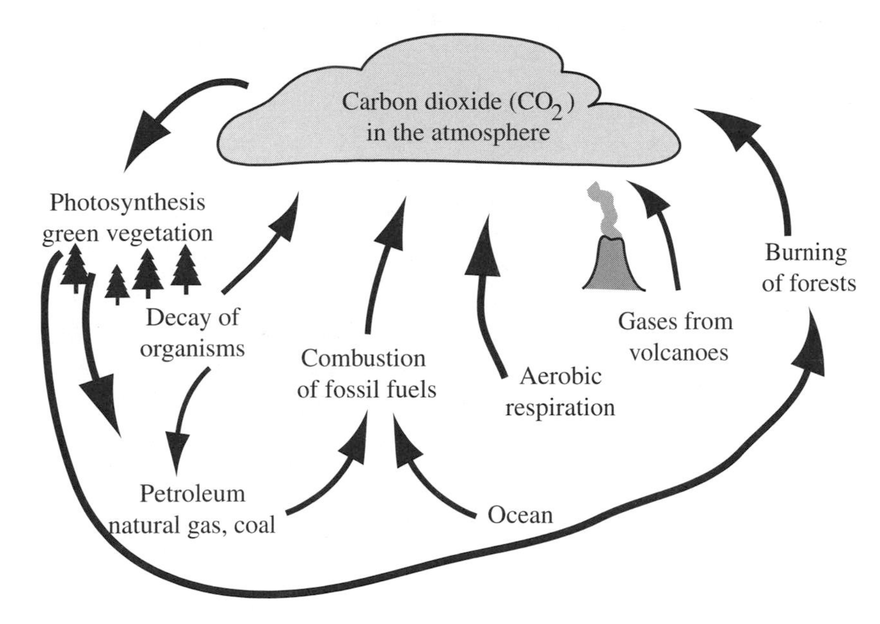

FIGURE 2-7 • Carbon Cycle The carbon cycle is a natural process in which carbon is cycled within the environment. Other than the water cycle, no mechanism is more crucial that the circulation of carbon between atmosphere, lithosphere, and the hydrosphere.

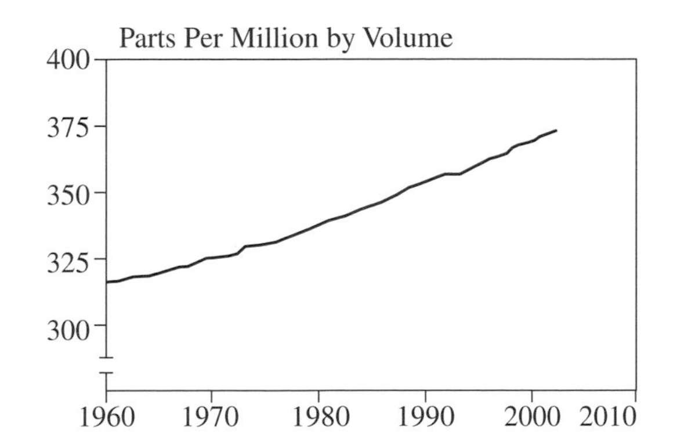

FIGURE 2-8 • Atmospheric Concentrations of Carbon Dioxide, 1960–2002
Source: Scripps Institute of Oceanography.

OTHER GREENHOUSE GASES

Other important greenhouse gases include nitrogen oxides (No_x), chlorofluorocarbons (CFCs), methane, and water vapor. As mentioned earlier, nitrogen oxides are caused by the burning of fossil fuels. Chlorofluorocarbons, or CFCs, also gradually enter the air from refrigerators and air conditioners. Methane is given off in fairly large amounts from agricultural activities, sanitary landfills, animal feedlots, and other sources.

Rising Global Temperatures

Global average temperatures have remained relatively stable over the last 10,000 years. But since 1880, when reliable temperature records started to be kept worldwide, the global average temperature has risen by nearly 0.5°C. In the last 120 years, snow cover in the Northern Hemisphere and floating ice in the Arctic Ocean have decreased. Globally, sea level has risen 10 to 25 centimeters (4 to 10 inches) over the past century.

Glaciers Are Disappearing

Environmentalists are now concerned that glaciers and ice sheets are disappearing too quickly because of warming conditions on Earth. Many scientists have pointed out that warming temperatures over the last 60 years have led to the drastic decline of glaciers. The glaciers of Glacier National Park, for example, like glaciers all over the world, are shrinking. Some of the park's glaciers have already shrunk by more than half, and the number of glaciers in the park has dropped from an estimated 150 in 1850 to approximately 50 in 1999. Scientists are concerned that the disappearing glaciers will have an adverse effect on the park's ecosystems. If nothing is done to cut down global warming, by the year 2030, park scientists predict, there may not be a single glacier left in Glacier National Park.

This area of the National Glacier Park in Montana was once covered entirely by a glacier. Since 1932 the glacier has all but disappeared. (Courtesy of Glacier National Park Archives, Jerry DeSanto, photographer)

Earth's northern latitudes have become about 10 percent greener since 1980, as a result of more vigorous plant growth associated with warmer temperatures and higher levels of atmospheric carbon dioxide.

The World Meteorological Organization announced that the 1990s was the warmest decade of the twentieth century. Eight of the 10 hottest years on record occurred in the 1990s. Other changes indicate that Earth is getting warmer:

- Glaciers are retreating or disappearing in parts of North and South America and Europe.
- Coral reefs are drying up as the ocean water temperatures are rising.
- The Arctic permafrost is beginning to melt.
- The movement of some migratory species such as polar bears, birds, and butterflies is being disrupted.

Effects of Climate Change

Climate change as a result of global warming is likely to have a significant impact on global environments. According to the UN Intergovernmental Panel on Climate Change (IPCC), climate models indicate that unless the world takes steps to reduce emissions of greenhouse gases, the global temperature is projected to rise another 0.7°–3.0°C (1°–4°F) by the year 2100. Other projections are higher. Just a few

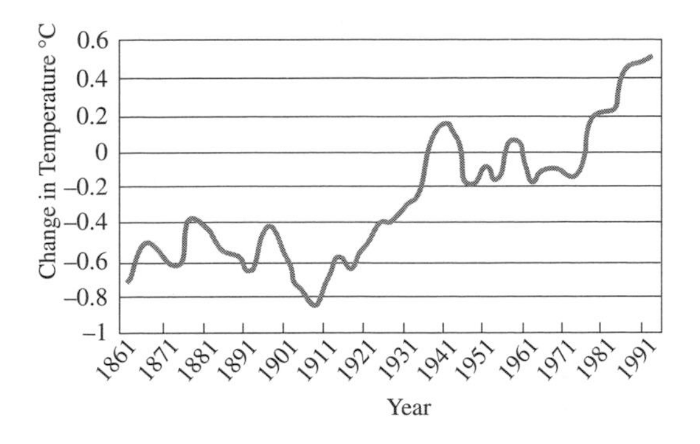

FIGURE 2-9 • Global Warming *Source:* United Nations Intergovernmental Panel on Climate Change, 1995.

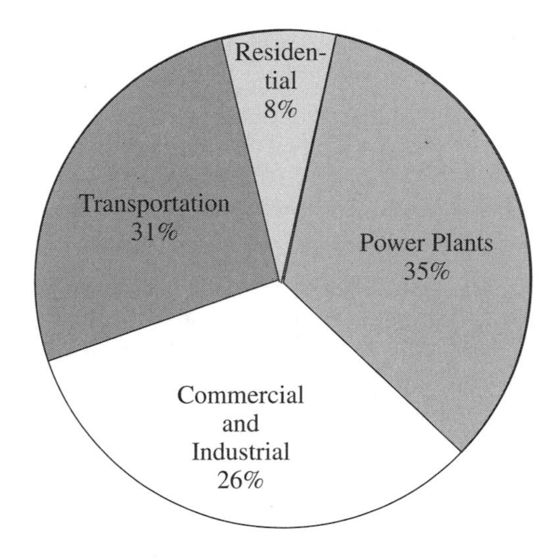

FIGURE 2-10 • Sources of Carbon Dioxide The combustion of fossil fuel and the burning of forests emits carbon dioxide into the atmosphere.

degrees of difference in the global average temperature can result in a major climate change.

At current emissions rates, mean sea level is projected to rise 15 to 95 centimeters (6 to 38 inches) by the year 2100. This would cause the flooding of low-lying areas and other damage particularly along the U.S. coast. Studies by the EPA and others estimate that along the Gulf and Atlantic coasts, a 30-centimeter (11-inch) rise is likely by 2050 and could occur as soon as 2025. Nationwide such a sea level rise could eliminate 17 to 43 percent of American wetlands, with more than half the loss taking place in Louisiana alone. Sea levels will likely continue to rise for several centuries even if global temperature increases are halted within the next few decades.

IMPACT ON THE ENVIRONMENT

Patterns of rainfall and snowfall are also expected to change, and many plant and animal species may not be able to adjust to such shifts. Climatic zones, ecosystems, and agriculture zones could extend 150 to 550 kilometers (90 to 340 miles) toward the poles in the mid-latitude regions. Human society will face new risks and pressures, including

Global Warming May Harm the Coelacanth

Scientists in Africa are concerned that a rise in ocean temperatures, as a result of global warming, will threaten a fish called the coelacanth. No ordinary fish, the coelacanth was believed to have been extinct for 70 million years. Then in 1938, one was caught by fishers in South African waters and was identified by a scientist. The catch made scientific history because the fish outlived the dinosaurs. Since then other coelacanths have been discovered living in colonies off Africa's east coast and Indonesia. In 2001, a colony was found living in the deep waters in Sodwana Bay, South Africa.

Much of the food the coelacanth depends on is derived from the nearby coral reefs. Any rise in sea temperatures would kill the coral reefs, which in turn would harm the coelacanth population. The fish would be forced to go deeper into the water where there is less food. Whether or not the coelacanth can adapt to the rising sea temperatures will be an ongoing story.

issues related to food security, water resources, economic activities, and human health. The changing climate could alter forests, crop yields, and the distribution of water supplies.

DID YOU KNOW?

The West Nile Virus was first discovered in Uganda in east-central Africa in the 1930s. The disease has traveled northward and is now a health concern in North America.

Global warming may also increase the prevalence of some infectious diseases, particularly those that appear only in warm areas. Diseases such as malaria and dengue fever, which are spread by insects, could become more prevalent if warmer temperatures were to enable those insects to extend their ranges. As an example, outbreaks of the West Nile Virus in the United States, caused by Culex mosquitoes, may be the result of increasing temperatures. The increased temperatures extend the range of insects that carry disease organisms.

PROBLEMS IN THE OZONE LAYER

Besides the climate changes and the global warming issues in the troposphere, the depletion of the ozone layer in the stratosphere is also a concern of environmentalists. Without the ozone layer, there would be little life on Earth.

The ozone layer in the upper atmosphere provides a protective layer that shields Earth from much of the sun's ultraviolet radiation (UV), which is dangerous to humans and ecosystems. Thus, the ozone layer protects organisms by preventing much of the radiation from reaching Earth.

One type of solar ultraviolet radiation, *UVB*, is harmful to the cells and tissues of organisms. It has been linked to skin cancers and eye cataracts in humans. Skin cancer is a problem for many people who like to sunbathe. Therefore, sun protection lotions are necessary for those who sunbathe as well as those who work outdoors. Unfortunately, UVB is also harmful to aquatic life and certain crops.

The ozone layer is located 10 to 50 kilometers (6 to 31 miles) above Earth's surface. Ozone forms when UV radiation splits an

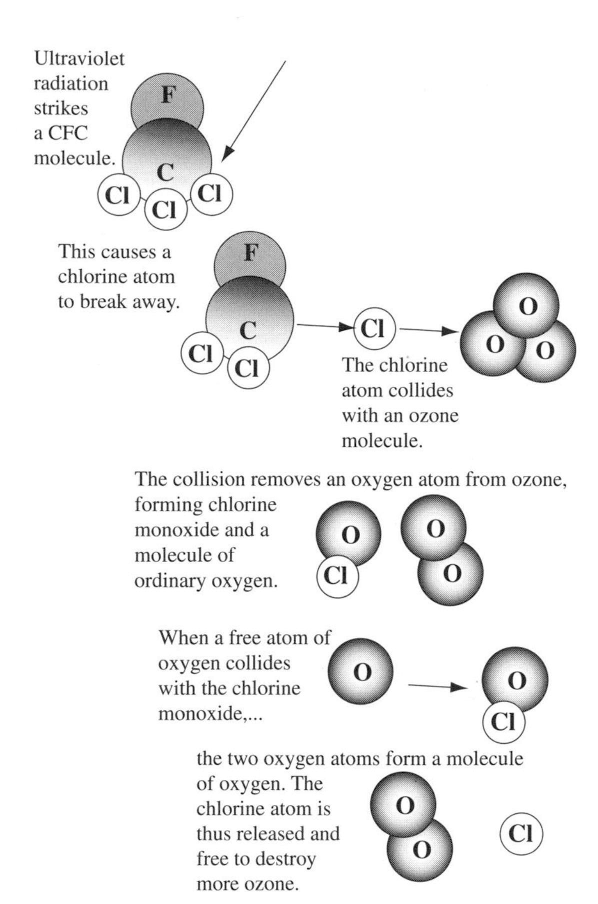

FIGURE 2-11 • Ozone Depletion Process
Source: NASA. 1993

oxygen (O_2) molecule into single oxygen atoms, which in turn collide with oxygen molecules to form ozone. The ozone (O_3) molecule also absorbs UV radiation. It breaks apart to form O_2 and a single atom of oxygen. The free oxygen atom then collides with an ozone molecule and splits into two oxygen molecules. The UV splits these molecules. These reactions occur repeatedly and simultaneously. Therefore, the total amount of ozone and oxygen in the stratosphere remains balanced as long as ozone forms and breaks apart at the same rate. However, the ozone layer is depleting. Why?

Ozone Depletion in Antarctica

In the 1980s, scientists at the British Antarctica Survey station located at Halley Bay, Antarctica, revealed that the extent of the ozone layer in Antarctica was decreasing in size. They discovered that each year more than half the natural ozone was being lost from the stratosphere over Antarctica following the seasonal return of sunlight. Research showed that the culprit was human-created air pollutants, particularly CFCs (chlorofluorocarbons), which were releasing more than a half-million tons of chlorine into the atmosphere every year, destroying thousands of ozone molecules.

In 1998 NASA satellites revealed that the area of ozone layer depletion over Antarctica was larger than ever, encompassing 26.9 million square kilometers (10 million square miles), 5 percent more than the previous record set in 1996.

Ozone Hole

The thinning and reduction in the area of an ozone layer is widely referred to as the ozone hole. The term "hole" is used in the popular press and other media when reporting on ozone depletion. However, the phenomenon is more correctly described as a low concentration of ozone.

The hole in the ozone layer occurs yearly in late September, which is the beginning of spring in the Southern Hemisphere. The hole is roughly the size of the continental United States and has increased in size each year since being discovered. Many scientists believe that low stratospheric temperatures lead to the formation of icy clouds. These clouds, which bring about chemical changes, result in the rapid ozone layer depletion during September and October.

The Total Ozone Mapping Spectrometer (TOMS) on board an Earth satellite has mapped the Antarctica ozone hole in detail. TOMS has also mapped the distribution of ozone over the entire globe. Ozone holes have appeared during the spring in Russia east of the Ural Mountains and in an area comprising the Baltic states, Moscow, and St. Petersburg. Ozone depletion is also occurring in the stratosphere over the Arctic, Panama, and New Zealand.

Health Risks

Increased ultraviolet radiation from ozone layer depletion may result in elevated skin cancer and cataract rates; suppression of the human immune system; disruption of plant life, including increased susceptibility to disease; reduction in phytoplankton growth; and eventual reduction of the numbers of aquatic species, including krill.

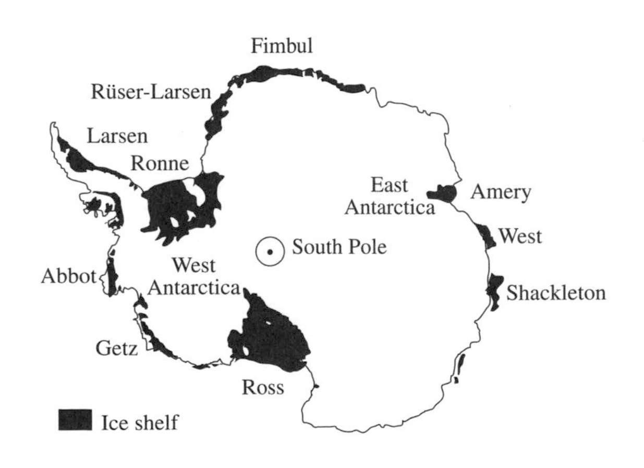

FIGURE 2-12 • Antarctica There are several ice shelves in the Antarctica. The major groups of Antarctic fauna are sea birds, penguins, seals, and whales. All Antarctic mammals are marine life, including seals and whales; the latter are protected in the Southern Ocean's International Whale Sanctuary.

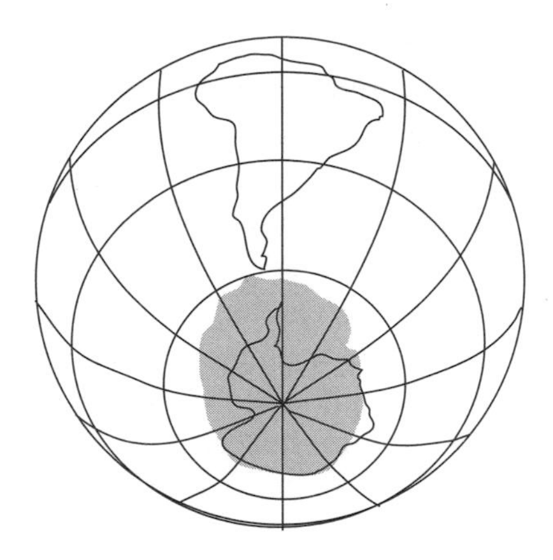

FIGURE 2-13 • Antarctic Ozone Hole, 1991
Before the1970s, there was no ozone hole over the Antarctica. By the early 1990s, the hole extended over the entire continent as shown in this diagram.

OZONE-DEPLETING CHEMICALS

The depletion of the ozone layer is caused by ozone-depleting substances (ODSs). These destructive chemical substances, many of them containing chlorine compounds, include CFSs, hydrochlorofluorocarbons (HCFCs), halons, methyl bromide, carbon tetrachloride, and methyl chloroform. Generally ODSs are very stable in the troposphere and only degrade under intense ultraviolet light in the stratosphere. When ozone-depleting substances break down, they release chlorine or bromine atoms, an element that has a structure similar to chlorine. These atoms deplete the ozone in the stratosphere.

MEASURING THE DAMAGE

A Dobson scale is used to measure the concentration of ozone levels and is a good indicator of ozone layer depletion. One Dobson unit indicates a concentration of one part per billion of ozone O_3. The measurement has been used worldwide for more than 30 years.

In the late 1950s, scientists measured the concentration of ozone in the stratosphere in Antarctica. At that time the concentration of ozone was approximately 300 Dobson units (DUs). However, by the 1980s the concentration of ozone had dropped to about 200 DUs. Since then, the concentration of ozone in Antarctica has reached a high of 250 DUs and a low of about 90 DUs. Seasonal variations and other natural conditions can produce large swings in ozone levels and changes in the concentration of ozone. Measurements in Leningrad, Russia, for example, have indicated ozone levels as high as 475 DUs and as low as 300 DUs.

PROTECTIVE MEASURES

Environmentalists have recommended a number of steps for reducing greenhouse gas emissions and for protecting the ozone layer. To reduce CO_2 emissions, they suggest substituting fossil fuels with cleaner energy sources such as solar energy, wind power, tidal power, and even nuclear energy. Conservation agencies are also looking at ways to reduce the burning of tropical rainforests, which adds billions of tons of CO_2 into the atmosphere each year.

The stabilization of atmospheric concentrations of greenhouse gases will demand a major international effort. The UN Framework Convention on Climate Change was adopted by many nations at the 1992 Earth Summit in Rio de Janeiro, Brazil. The main objective of the convention was to consider ways to stabilize the concentrations of greenhouse gases in the atmosphere. At another meeting of the nations in Kyoto, Japan, in 1997, the countries agreed to a reduction of greenhouse gas emissions below 1990 levels of approximately 5 percent by 2010. The agreement reached at this meeting is known as the Kyoto Protocol. The Kyoto Protocol had to be ratified by at least 55 countries before it could be enforced. The 55 nations must account for at least 55 percent of the total 1990 CO_2 emissions of developed countries. As of September 1998, more than 55 countries had signed the Kyoto Protocol. The Kyoto Protocol represents a start on the cooperation of developing countries to reduce emissions.

Protecting the Ozone Layer

A 1987 international treaty, called the *Montreal Protocol*, was approved by 30 countries for phasing out the production of CFCs that deplete the ozone layer in the atmosphere. Since that time, members of the Montreal Protocol are forbidden to purchase CFCs or products containing them from nations that have not agreed to the treaty.

Without the Montreal Protocol and its amendments, the continuing use of CFCs and other compounds would have tripled the stratospheric abundances of chlorine and bromine by the year 2050. Two other harmful chemicals, halons and carbon tetrachloride, were phased out in 2000. Methyl chloroform will be phased out in 2005. The phasing out of these chemicals was based on the scientific theory that, these substances could once emitted into the atmosphere, significantly deplete the stratospheric ozone layer that shields the planet from damaging UVB radiation. The Montreal Protocol agreement has also addressed the use of HCFCs, the replacement chemical for CFCs. Although HCFCs are less harmful to the ozone layer than CFCs, they will be banned by the year 2040 or earlier.

Vocabulary

Biogeochemical cycles The flow of chemical substances to and from the atmosphere, hydrosphere, lithosphere, and living organisms. The nitrogen cycle, the carbon cycle, and the oxygen cycle are major biogeochemical cycles.

Climate Weather conditions in a particular place over a period of time.

Leach or Leaching A process by which substances or materials are washed out of the soil. Fertilizers, for example, can leach out of the soil and be carried to nearby waterways.

Montreal Protocol An international agreement signed in 1987 to stop the production of chemicals that deplete the ozone layer.

Ozone An oxygen molecule composed of three oxygen atoms (O_3), which forms when oxygen (O_2) gas in the atmosphere is exposed to ultraviolet (UV) radiation.

Parts per million (ppm) Units of measure commonly used to describe the concentration of a toxic chemical in another material, such as pesticides in a food product. The units can be used to establish the maximum permissible amount of a contaminant in water, land, or air. (May also describe naturally occurring concentrations.)

UVB A classification of solar ultraviolet radiation that is dangerous and can cause skin cancer.

Activities for Students

1. Using the tools from a chemical lab and the principles of condensation and evaporation, create and conduct an experiment that purifies air and water by separating them from contaminants.
2. Acquire litmus paper and conduct tests on various water samples in your school, home, and town. Test the results on the various liquids that we consume such as soda, juice, and water.
3. Study the wind currents in and around the midwestern United States and their paths across the northeast section of the United States to infer how air pollutants, such as acid rain particulates, can travel.

Books and Other Reading Materials

Edgerton, Lynn T. *The Rising Tide: Global Warming and World Sea Levels.* Washington, D.C.: Island Press, 1991.

Forster, Bruce. *The Acid Rain Debate: Science and Special Interest in Policy Formation.* Natural Resources and Environmental Policy Series. Ames: Iowa State University Press, 1993.

Global Climate Change Digest: A Guide to Current Information on Greenhouse Gases and Ozone Depletion, edited by Dr. Robert Pratt for Enviornmental Information, 1988–. Monthly Abstracts.

Huttermann, Aloys, and Douglas, Godbold eds. *Effects of Acid Rain on Forest Processes.* Wiley Series in Ecological and Applied Microbiology. New York: John Wiley & Sons, 1994.

Lyman, Francesca, Irving Mintzer, Kathleen Courrier, and James MacKenzie. *The Greenhouse Trap: What We're Doing to the Atmosphere and How We Can Slow Global Warming.* A World Resources Institute Guide to the Environment. Boston: Beacon Press, 1990.

Peters, R.L. "The Effect of Global Climatic Change on Natural Communities." In E.O. Wilson, ed., *Biodiversity* (pages 450–461). Washington, D.C.: National Academy Press, 1988.

Roan, Sharon. *Ozone Crisis: The Fifteen-Year Evolution of a Sudden Global Emergency.* New York: John Wiley & Sons, 1989.

Schneider, Stephen. *Global Warming: Are We Entering the Greenhouse Century?* New York: Random House, 1989.

The United Nations Environment Programme (UNEP) has prepared a *Montreal Protocol Handbook* that provides additional detail and explanation of the provisions. CIESIN's *Thematic Guide on Ozone Depletion and Global Environmental Change* presents an-in-depth look at causes, human and environmental effects, and policy responses to stratospheric ozone depletion.

Websites

Environmental Protection Agency, http://www.epa.gov/docs/acid rain/andhome/html

Environmental Protection Agency, Acid Rain Program, http://www.epa.gov/docs/acidrain/effects/enveffct.html USGS Water Science, Acid Rain, http://www.ga.usgs.gov/edu/acidrain.html

Environmental Protection Agency Global Warming Site, http://www.epa.gov/global warming

Greenpeace International—Climate, http://www.greenpeace.org/~climate

UN Intergovernmental Panel on Climate Change, http://www.ipcc.ch

United Nations Environment Programme, http://www.ipcc.ch

CHAPTER 3

Water Pollution on Land and Sea

In 2003, the world's population reached more than 6 billion people and continues to grow in numbers. At the same time, the amount of freshwater per person is shrinking. By 2015, approximately 3 billion people will live in countries where getting enough water for basic needs will be difficult. More than 1 billion people will lack access to clean drinking water; others will die from contaminated water.

Oceans, estuaries, salt marshes, mangrove forests, and coral reefs are also not safe from pollution. Land pollution, some of which includes chemicals, heavy metals, garbage, and pesticides, is washed away by rivers and streams into coastal-lying areas and into the ocean. The dumping of wastes into the ocean destroys marine ecosystems. This chapter considers some of the freshwater and saltwater pollution problems we have today.

FRESHWATER POLLUTION

Refer to Volume II for more information about freshwater uses.

DID YOU KNOW?

If all of the world's water were poured into a gallon jug, the amount of freshwater available would be about a tablespoon.

In the United States, 218 million people live within 18 kilometers (10 miles) of a polluted river, lake, stream, or coastal area. The U.S. Environmental Protection Agency (EPA) estimates that at least a half-million cases of illnesses annually can be attributed to contaminated drinking water in the United States.

Comprising over 70 percent of Earth's surface, water is undoubtedly the most precious natural resource that exists on our planet. Most living organisms depend on water for their existent. However, water pollution and the overuse and depletion of freshwater sources are creating a global water crisis.

Potable Water

Water that is safe for people to drink is called potable water. Potable water must be fresh (have a salinity lower than 0.35 grams per liter) and free of contamination by pollutants or *pathogens*. Potable water is generally free of minerals that give the water a bad taste or produce an objectionable odor, and it is neither too acidic nor too *alkaline*.

The demand for potable water often exceeds the supply in remote areas and developing nations. In such places, people usually rely on

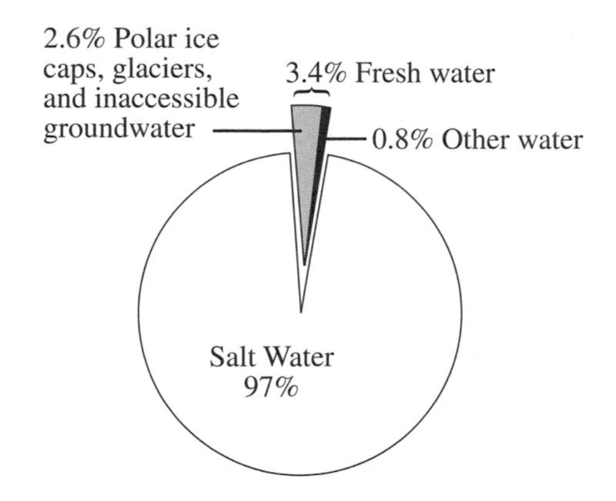

FIGURE 3-1 • Freshwater Sources Approximately only 3 to 3.5 percent of all water on Earth is fresh water. However, most of the freshwater is stored in ice caps and glaciers, which is inaccessible for human needs.

TABLE 3-1 Water Used in the Home

Task	Water Used (liters)
Showering for five minutes	95.0
Brushing teeth	10.0
Washing hands	7.5
Flushing standard toilet	23.0
Flushing "low-flow" toilet	6.0
Washing one load of laundry	151.0
Running dishwasher	19.0
Washing dishes by hand	114.0

surface water for their water supplies. This water may be shared with wildlife that use the water as habitat, for drinking, or for cooling. It may be contaminated with pathogens or be unclean because of suspended particulates or algae. Despite these problems, people are forced to use such water for drinking, cooking, and bathing because it is the only available water.

The degradation of water quality is called water pollution. Water can be polluted by excessive amounts of chemicals such as *heavy metals*, organic chemicals, disease-causing microorganisms, and sediments, as well as by heat or thermal pollution. These pollutants come from homes, mining activities, municipal sewage plants, industry and manufacturing plants, ranching, and agricultural operations.

Agricultural Pollution

The release of pollutants derived from farming or ranching practices into the environment is called agricultural pollution. Agricultural pollution may result from the use of agrochemicals, such as fertilizers and pesticides; through the runoff of wastes and nutrients from feedlots or

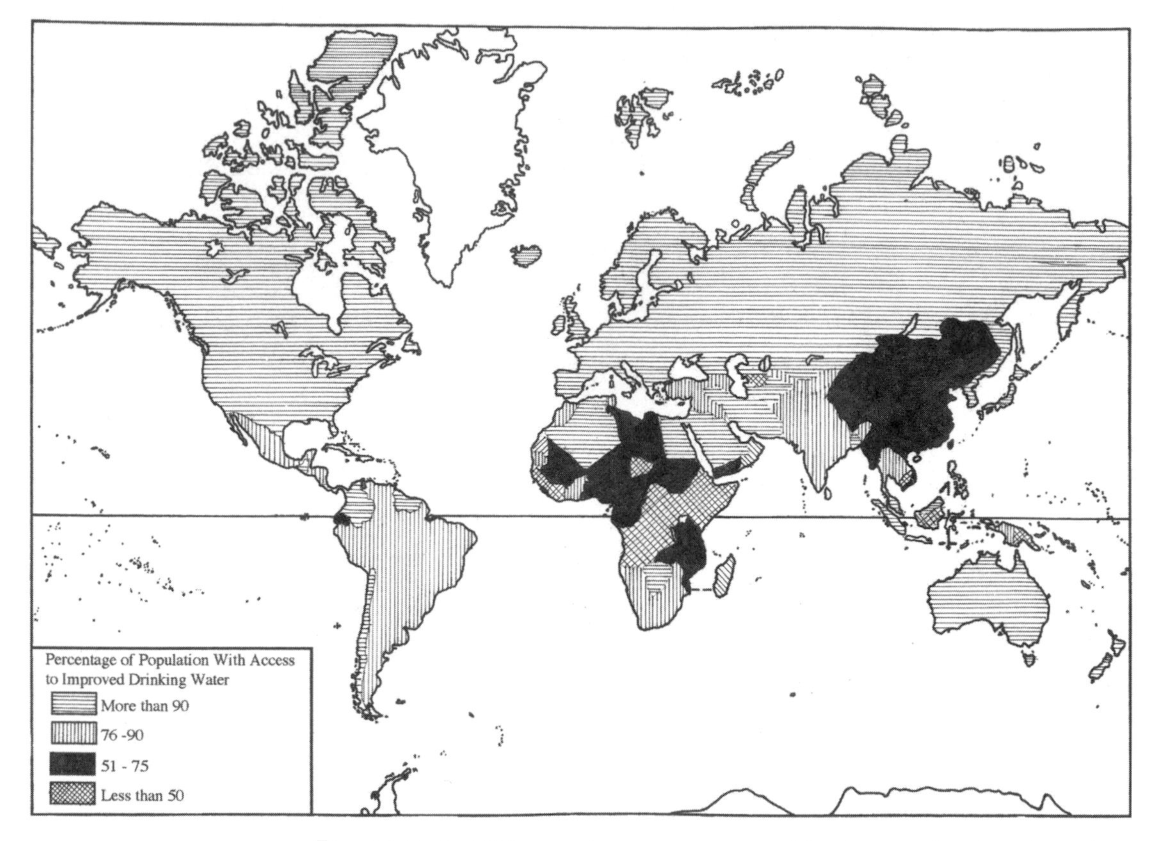

FIGURE 3-2 • In 2003, researchers reported that as Africa's population rises, demand for household water is projected to grow faster than anywhere else on the planet. This means that up to 523 million people will be without access to clean water by 2025 unless governments invest in better infrastructure to provide fresh water to the populations.

Medicines, Household Cleaners, and Beauty Aids Are Contaminating Waterways

Scientists are now studying another possible source of contaminants found in rivers and streams that are sources for municipal freshwater supplies. A government study reported that the nation's rivers and streams contain traces of household chemicals. These chemicals are used in the production of medications, insect sprays, household cleaners, beauty aids, nicotine, and foods. The products are sold on supermarket shelves and are found in most households, as well as on farms and in factories. Wastes from these products find their way into rivers and streams that are used for drinking water.

Scientists call this new class of contaminants pharmaceutical and personal care pollutants, or PPCPs. Scientists report that there are no safety laws in regulating these contaminants at municipal wastewater treatment plants. For now, the long-term effects of exposure to these kinds of pollutants are unclear, but water utility officials are concerned that these pollutants can be harmful. They want the EPA to decide how to regulate PPCPs. The water utility officials believe that promising new wastewater treatment technologies can break down many of the chemicals through biological methods.

TABLE 3-2 Water Pollutants

Point Sources	Bacteria	Nutrients	Total Dissolved Ammonia	Solids	Acids	Toxics
Municipal sewage treatment plants	•	•	•			•
Industrial facilities				•		•
Combined sewer overflows	•	•	•			•
Nonpoint Sources						
Agricultural runoff	•	•		•		•
Urban runoff	•	•		•		•
Construction runoff		•				•
Mining runoff				•	•	•
Septic systems	•	•				•
Landfills spills						•
Forestry runoff		•				•

Source: U.S. Environmental Protection Agency.

ranches; and through unsustainable agricultural practices that cause topsoil erosion.

FERTILIZERS

Chemical fertilizers are added to the soil to replace nutrients and to promote plant growth. However, when fertilizers are washed away into aquatic ecosystems, eutrophication occurs. Eutrophication happens when too many nutrients are present in an aquatic ecosystem. The excessive amounts of nutrients cause a population explosion of algae, called an algal bloom. During an algal bloom, the algae population can exceed the carrying capacity of the ecosystem and die. In time, the growing algae and plants use up the nutrients in the water. Once the food source is depleted, they begin to die in great numbers. In turn, bacteria and fungi begin the process of decomposition and release large amounts of organic matter into the ecosystem. At the same time, the decomposers use up much of the *dissolved oxygen* in the water, making the oxygen unavailable to other larger organisms, such as fish. Unable to meet their oxygen needs, these organisms also die, and the ecosystem becomes unable to support aquatic organisms.

SEDIMENTS

Excessive amounts of sediments produced another kind of water pollution. Sediments are tiny particles and flakes of sand, silt, and clay. When the land surrounding aquatic ecosystems is cleared of vegetation for agricultural activities, the construction of roads, logging operations, mining, and the building of houses and shopping centers, the soil becomes more vulnerable to erosion. The eroded soil is carried away into nearby waterways, where the sediments can adversely affect the organisms living there. Most of the heavier sediments sink to the bottom (sedimentation). The sediments and particles that sink to the bottom can smother fish eggs, thereby reducing fish populations. If these populations decline, other organisms that depend on them for food will also be affected.

Lighter particles, particularly silt and clay, are suspended in water. These floating particles can become so dense that they can prevent sunlight from penetrating the water, a phenomenon known as "turbidity." Without sunlight, aquatic organisms that need photosynthesis to live, such as algae, phytoplankton, and plants, may not be able to make enough food for their survival.

Thermal Pollution

In many countries, industries and factories use large amounts of cooling water for their machinery. Electric power plants, for example, account for about 85 percent of all water used in cooling machinery. Once the cool water is used, some industries and power plants discharge the heated water into cooling towers and ponds. These systems allow the discharged water to cool off before being released into nearby waterways.

However, thermal pollution can occur if the discharged water is released directly into nearby streams, rivers, or lakes. The heated water increases the water temperature. As the temperature rises, the amount of dissolved oxygen in the water decreases. A drop in dissolved oxygen levels can disrupt ecosystems and prevent fish from spawning.

WATER-BORNE DISEASES

More than 5 million people die each year from water-borne diseases such as chlorea and typhoid. Many of these people live in Asia and Africa, and most of them are under the age of five.

The major water-borne pathogens include giardia, cryptosporidium, and coliform bacteria, among others. Giardiasis is an intestinal disease caused by ingesting water containing *Giardia lamblia*. Giardia is a protozoan contaminant in the feces of humans and animals. Ingestion of this protozoan in contaminated drinking water can cause severe stomach and intestinal diseases that may last for weeks or months. Symptoms of the disease include diarrhea, fatigue, and stomach cramps. Unless the drinking water is treated, giardia can live in contaminated water from one to three months.

Cryptosporidium is a protozoan pathogen that is associated with the disease cryptosporidiosis, which can be transmitted by contaminated drinking water. This disease can cause diarrhea, stomach cramps, vomiting, and fever. The disease can be fatal in humans who have weak immune systems.

The largest outbreak of cryptosporidiosis in the United States took place in Milwaukee, Wisconsin, in 1993, according to the EPA. Milwaukee's water supply comes from Lake Michigan and is treated and filtered by a treatment plant before being distributed to homes and businesses. During 1993 there was a period of heavy rains and runoffs, making the water treatment plant ineffective. As a result, more than 400,000 people were affected by the disease, and 4,000 needed to be hospitalized. The disease caused over 50 deaths. The original source of the contamination remains uncertain.

Coliform Bacteria

Coliform bacteria microorganisms present in the intestinal tracts of humans and other vertebrates become an environmental problem when discharged into bodies of water, usually sewage, or when they contaminate food. Coliform bacteria such as *Escherichia coli* live within the intestinal tracts of humans and animals, such as cattle and horses. In the digestive tract, these bacteria aid in the digestive process of the host by breaking down cellulose or other matter that the host cannot digest. However, when these bacteria are ingested from contaminated water or food, they become pathogens. In other words, they can cause acute disease and possibly death for the organisms in which they normally live.

The presence of coliform bacteria in water is evidence of pollution. Such pollution is common in some developing nations where potable water is in short supply. People obtain drinking water from sources exposed to farm runoff, are used by animals as drinking and cooling ponds, or are used for bathing by humans. To help prevent the spread of *fecal coliform* in the United States, drinking water supplies are routinely tested and assigned a coliform index, a rating of water purity based on a count of fecal bacteria.

Illness caused by coliform bacteria can result from ingesting food contaminated with these bacteria. Food becomes contaminated through poor hygiene practices of humans and through poor handling practices at meat-slaughtering and -processing plants. In the United States, inspectors working for the U.S. Food and Drug Administration (FDA) and state and city health departments are responsible for examining meat products, meat-processing facilities, and restaurants to prevent the spread of bacteria in foods. This effort is limited, however, and so does not protect the entire food supply from contamination. To help offset this problem, restaurants serving meats that could be contaminated with coliform bacteria are encouraged to cook such foods until their temperatures reach 68.3°C (155°F) to kill any bacteria that may be present. Individuals are encouraged to do the same in their homes or to cook meats until the juices run clear.

TRACKING THE SOURCES OF WATER POLLUTION

Point Sources

Water pollution comes from two basic sources or types—point sources and nonpoint sources. Point sources are places where harmful substances are emitted directly into a body of water. These sources are stationary locations where pollutants are discharged into the water from pipes, wells, ditches, tunnels, or sewers and are easily identifiable. Some common point sources of pollution include discharges from sewage treatment plants, factories, power plants, and mining operations.

Other point sources of water pollution include stormwater outfalls and agricultural feedlots. Thermal or heat pollution can result from discharge of runoff from hot pavement or from industrial cooling water. Many of these sources can be regulated by controlling drainage, by treating wastewater before discharge, or by eliminating the discharge altogether. Point sources are at specific locations that can be mapped as single points where municipal sewer treatment plants and industrial facilities discharge wastes directly into surface water.

Nonpoint Sources

Water pollution from nonpoint sources is a major factor in the pollution of streams and rivers. Nonpoint sources consist of large areas containing pollutants that enter water at many different locations within a *watershed*. They do not come from clearly identifiable places. Estimates indicate that nonpoint source pollution accounts for more than 70 percent of all U.S. surface water pollution, according to the Izaak

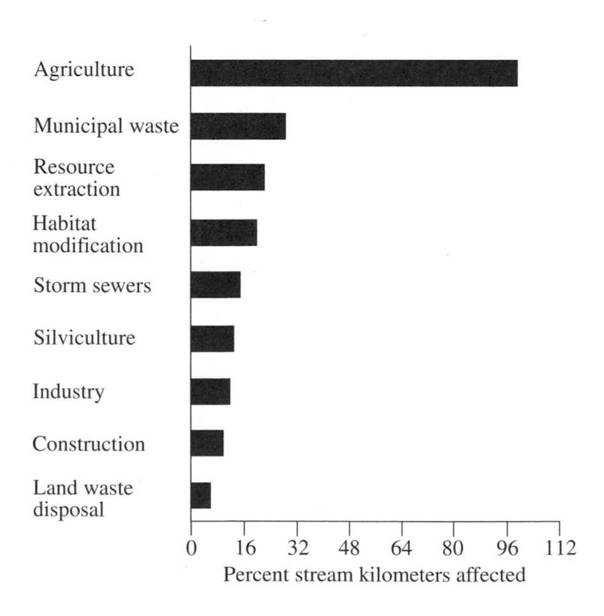

FIGURE 3-3 • United States Nonpoint Source Pollution Nonpoint sources of water pollution is a major factor in the pollution of steams and rivers. (Courtesy of United States Environmental Protection Agency, 1990)

Walton League. Nonpoint sources are much more difficult to control than point sources.

Nonpoint sources include acid rain, runoffs from highway construction, paved urban areas, landscaped areas such as golf courses, mining and construction activities, and forestry. However, agricultural runoff is the major concern for water pollution, since the usage of pesticides and fertilizers by many developed regions in the world is growing. Agricultural runoff also includes manure from livestock such as dairy cows. The pollution occurs because the soils in the watershed cannot absorb and filter all the runoff of organic and inorganic pollutants. As a result, high concentrations of pesticides, nitrates, and phosphorus materials as well as soil erosion can enter waterways and cover streambeds. This action decreases the oxygen content of the water and reduces the ability of plants and animals to support themselves. These pollutants impact water quality and can threaten drinking water supplies.

Today nonpoint pollution from agricultural and urban runoff remains the largest source of water quality problems. Nonpoint pollution is the main reason why approximately 40 percent of rivers, lakes, and estuaries in the United States are not clean enough to meet the standards for basic uses such as fishing or swimming.

GROUNDWATER POLLUTION AND DEPLETION

Not all freshwater is found at the surface level. Globally, about 30 percent of all freshwater is underground. In this country, approximately 50 percent of the people depend on underground water stored in *aquifers*. In fact, many of the rural areas of the United States depend almost entirely on groundwater.

Aquifers store large quantities of water below the ground where the water is usually better protected from the effects of surface contaminants. However, aquifers require protection and careful monitoring to ensure that they do not become polluted by groundwater contamination as a result of inappropriate land uses or accidental chemical leaks or spills or depletion by excessive withdrawal.

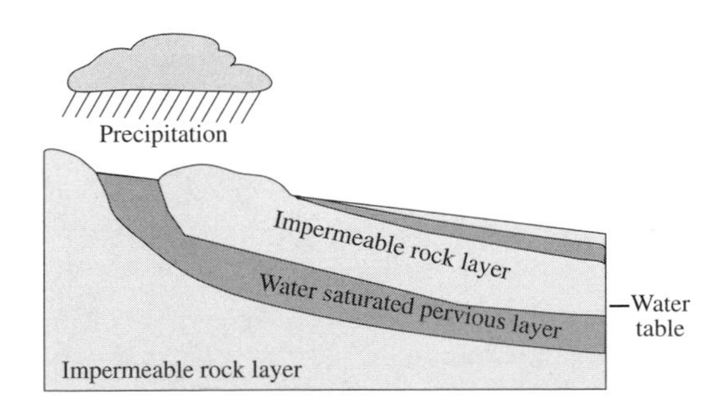

FIGURE 3-4 • Aquifer An aquifer is a permeable underground layer through which underground water flows relatively easily. Underground water resources are an important source for drinking water and irrigation in most parts of the world.

DID YOU KNOW?

In India, millions of people suffer from arsenic poisoning from arsenic found in drinking water. Many years ago, shallow wells were dug into aquifers for freshwater uses. However, the naturally occurring arsenic in the soil made the water toxic to humans. Scientists are now developing strategies to clean up the arsenic, a carcinogen, in the water.

Aquifers are being polluted by nutrients, heavy metals, pesticides, petroleum products, and other hazardous chemicals. Leaking from undetected underground storage fuel tanks is a special contributor to the problem. An EPA study found more than 300,000 underground tanks that are leaking pollutants into aquifers. Water officials in several states have reported that a gasoline additive called MTBE was leaking from some gas stations' underground fuel storage tanks. MTBE is an additive in gasoline that was developed by petroleum companies in the 1970s to replace lead additives and reduce automobile exhaust emissions. The additive has done a excellent job in keeping the air clean. However, MTBE, which is a very soluble, has leached into groundwater throughout much of the United States. If the additive escapes from broken underground tanks, it can contaminate groundwater used for drinking. As a result of the MTBE contamination, several public drinking wells have been closed. MTBE has been banned in California of 2000 and in New York as of 2004. However, environmentalists and water quality experts want a national ban on the use of MTBE.

Aquifer Depletion

Aquifers are threatened not only by pollution but also by depletion, which is becoming a major crisis. Depletion occurs when the amount of water withdrawn (pumped) from an aquifer is greater than the amount of water entering (recharging) the aquifer. If too much water is withdrawn in a large aquifer, the ground surrounding it can sink. In Mexico City, for example, the underground water table has dropped 13 to 16 meters (40 to 50 feet) in some places. As a result, some sections of the city are sinking.

TABLE 3-3 Threats to Estuarine Ecosystems

Pollutant	Source	Effects
Nutrients	Fertilizers; sewage	Algal blooms
Chlorinated hydrocarbons pesticides, DDT, PCBs	Agricultural runoff; industrial waste	Contaminated and diseased fish and shellfish
Petroleum hydrocarbons	Oil spills; industrial waste; urban runoff	Ecosystem destruction
Heavy metals (arsenic, cadmium copper, lead, zinc)	Industrial waste; mining	Diseased and contaminated fish
Particulate matter	Soil erosion;dying algae	Smothers shellfish beds; blocks light needed by marine plant life
Plastics	Ship dumping; household waste; litter	Strangles, mutilates wildlife

OGALLALA AQUIFER

In the United States, some environmentalist believe that one of the largest aquifers in North America will be completely depleted by 2020 because of the amount of water being withdrawn. The name of the aquifer is the Ogallala.

The Ogallala Aquifer is located in the midwestern United States beneath portions of South Dakota, Nebraska, Colorado, Kansas, Oklahoma, New Mexico, Wyoming, and northern Texas. The quantity of water stored within the Ogallala Aquifer is estimated to equal the volume of Lake Huron, one of the Great Lakes. This great groundwater resource provides irrigation water for approximately 4,005,000 hectares (10,000,000 million acres) of farmland. The groundwater level beneath the Ogallala Aquifer has been dropping since the 1940s.

In parts of Texas, Kansas, and New Mexico, water levels in the aquifer have declined by more than 30 meters (90 feet) since water-pumping operations started in the 1940s. Water levels in the aquifer underlying Oklahoma have also declined. This indicates that the aquifer is shrinking and not being renewed. As a result of the depletion of water in the aquifer, farming areas in parts of Texas and New Mexico have cut back on using the aquifer for irrigation purposes. The water tables of aquifers throughout the world are also falling, and experts warn that the situation is expected to get worse as a result of the growing global population.

Refer to Volume II for more information on aquifers and groundwater.

DID YOU KNOW?

Humans use almost 50 times as much water today as they did 300 years ago.

FIGURE 3-5 • Ogallala Aquifer The Ogallala aquifer in mid-Western United States is the largest aquifer and groundwater source in North America. Many farmers and households in this section of the country rely on groundwater for their fresh water needs.

Saltwater Intrusion

Freshwater aquifers can also be threatened by a phenomenon called saltwater intrusion. Saltwater intrusion occurs in coastal areas when saltwater moves into a fresh aquifer. Saltwater intrusion sometimes becomes a problem when saltwater is drawn into water supply wells that pump groundwater from coastal aquifers. If too much fresh water is pumped out, the saltwater rises in the aquifer and intrudes farther inland, contaminating drinking water. Places where saltwater intrusion occurs include Cape Cod and Long Island in the northeastern United States.

CLEANUP OF WATER POLLUTION

Clean Water Act

A major federal law that regulates water pollution is the Clean Water Act (CWA). The Clean Water Act and other regulations protect surface water from pollution by controlling and regulating discharges of wastewater, stormwater, and dredge and fill materials. Other goals of the Clean Water Act include the elimination of discharges from chemicals and the maintenance of water quality that is suitable for fishing and swimming. The Clean Water Act also established restrictions on wastewater discharges and requirements for prevention and the cleanup of oil spills. It also regulates shellfish harvesting.

Sewage Treatment Plant

The Clean Water Act focuses on improving sewage treatment plants. A sewage treatment plant is a facility designed for the collection, treatment, and sanitary disposal of sewage (wastewater) from municipal and/or industrial sources. A major role of a sewage treatment plant is to separate solids, semi-solids, and other dissolved substances carried in wastewater from the water. The plant then cleans and purifies the water and releases the water back into the environment to be reused by people and other organisms.

Most municipalities in industrialized nations have some types of wastewater disposal procedures that make use of sewage treatment plants. Methods used for reclaiming water from sewage include the removal of solids from wastewater by screening, filtering, sedimentation, flotation, chemical coagulation, and flocculation. These processes are known as primary treatment. Next, the sewage is acted on by bacteria and other microorganisms that convert organic matter contained in the sewage into such compounds as water, carbon dioxide, nitrates, phosphates, and other organic and inorganic matter. This process is known as secondary treatment. Following secondary treatment, the remaining water, which may contain microbes, may be chemically treated to remove any phosphorus and dissolved solids still in the water. In addition, the water may also be treated with chlorine or ozone to kill any remaining potential pathogens. This phase of treatment, known as tertiary treatment, purifies the water, rendering it essentially free of microbes.

Safe Drinking Act

The Safe Drinking Act is a federal law enacted in 1974 in response to outbreaks of water-borne disease and increasing chemical contamination of public drinking water in the United States. This law focuses on the protection of all waters that are suitable for public drinking use from both above-ground and underground sources. The act protects drinking water supplies by establishing water quality standards for drinking water, monitoring public water systems, and guarding against groundwater contamination from injection wells. To date the EPA has identified enforceable standards for 80 contaminants, including organic compounds such as benzene; certain bacteria, viruses, and protists such as cryptosporidium; and inorganics such as lead, mercury, and copper.

Treating Drinking Water

Municipalities and towns are responsible for making sure that their water is safe to drink. The treatment of freshwater from reservoirs, lakes, rivers, and aquifers is processed by large cities at public treatment plants. The typical water treatment process includes several steps to remove leaves, trash, mud, algae, and disease-causing organisms. Chlorination is used to kill organisms, and samples of the drinking water are constantly tested to assess water quality before distribution the processed water to homes and businesses.

Refer to Volume II for more information on how drinking water is treated.

In developed countries like the United States, water supplies are often provided to homes and businesses through a community or city water company. In such cases, water potability is usually monitored at water treatment plants. However, accidents or natural disasters that allow harmful substances to enter the water supply can still occur. For example, flooding following severe storms can damage parts of water transport systems, allowing pathogens from sewage or other sources to enter

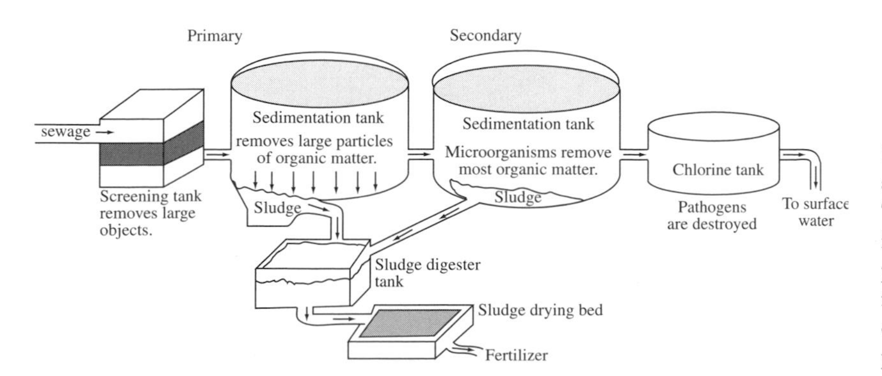

FIGURE 3-6 • Primary and Secondary Waste Treatment The illustration shows the process for treating wastes from removing large objects to destroying pathogens.

the water supply. When such problems arise, people may be instructed to boil water prior to use or to purchase bottled water until the system can be flushed, cleaned, and returned to proper working order.

Water Conservation

A process that is aimed to reduce unnecessary water usage or minimize the loss of water through waste is called water conservation. Types of industrial water conservation include the use of more efficient manufacturing processes, or "closed-loop" systems that make use of recycled water. Effective wastewater treatment systems can help conserve water by making water suitable for reuse.

Other water conservation measures include

- upgrading the old water supply piping systems by repairing leaks,
- restricting seasonal water usage through local water bans,
- implementing water-saving agricultural techniques, such as drip irrigation, that reduce water losses from evaporation, and
- implimenting designs that enhance recharge (reduce runoff).

The uses of low-flow plumbing devices in faucets, shower heads, and toilets are additional examples of how to conserve water. Water conservation is a critical issue because the rate at which water consumption is increasing threatens to deplete available freshwater resources. In the United States, water usage and water conservation are often measured in gallons (or liters) of water per person per day.

OCEAN WATER POLLUTION

Most ocean pollution comes from land-based sources. Runoff from farms, lawns, cities, and towns is the most significant source of ocean pollution. When pollutants such as industrial wastes, sewage, discarded oil from cars, pesticides, and fertilizers enter rivers as runoff, the rivers carry the wastes into the ocean. Coastal development such as mining and the construction of homes, hotels, dams, and canals is a major source of land-based ocean pollution. Dredging of ports and channels also adds to the problem. Coastal development not only contributes to runoff pollution but also destroys habitat and breeding grounds for fish, shellfish, and other aquatic organisms.

For years, oceans have been a favorite dumping ground for human activity. Each year, millions of tons of plastic six-pack rings, garbage bags, nylon fishing nets, fishing lines, construction materials, and other forms of garbage are dumped illegally into the oceans. Many animals die by becoming entangled in nets and strangling or suffocating to death.

Plastic wastes are a significant ocean pollutant because they do not break down easily. Turtles and other animals may also eat clear plastic bags that resemble jellyfish and die from suffocation or blockage of their digestive systems. The U.S. Office of Technology Assessment estimates that discarded plastic alone kills more than 1 million birds and more than 100,000 seals, sea lions, otters, whales, dolphins, sharks, and turtles annually.

Polluted Ocean Environments

Ocean pollution is causing the destruction of coral reefs, mangroves, and salt marshes.

Coral Reefs

Coral reefs are dying faster than previously estimated, according to a study by the United Nations. The world's reefs are spread among 101 countries and territories. A study released in 1999 by the World Resources Institute stated that nearly 60 percent of Earth's living coral reefs are threatened by human activity, including coastal development, overfishing, and inland pollution. Sewage and fertilizer runoff breeds algae that crowd out and smother coral. Global warming is also believed to have damaged reefs in recent years by causing bleaching—a loss of color that occurs when a reef's surface coating of algae dies.

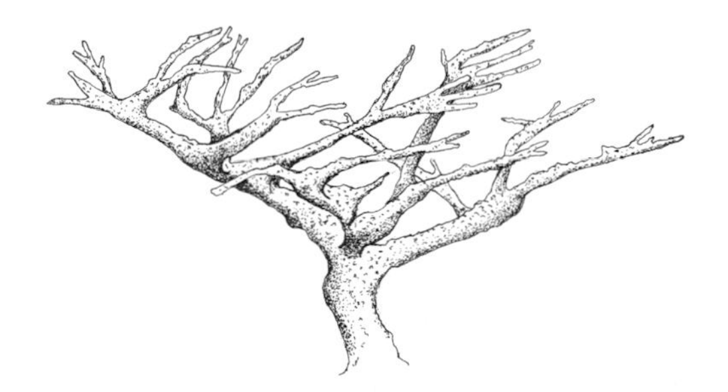

Figure 3-7 • The Elkhorn coral is an example of one type of coral. The loss of coral habitats has alarmed environmentalists because it is one of the world's critical ecosystems.

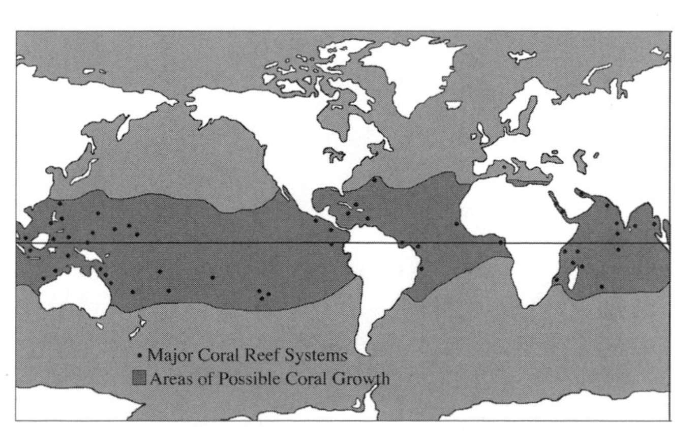

Figure 3-8 • **Coral Reefs** The world's largest coral site is Australia's Great Barrier Reef, which extends more than 1,600 kilometers along the northeast coast of Australia.

Healthy coral live in warm ocean water with sufficient light. (Courtesy of Craig Quirolo)

Coral can be destroyed due to coral bleaching, the result of an increase in warm ocean water temperatures. (Courtesy of Craig Quirolo)

MANGROVES

Estimates indicate that more than 50 percent of the world's mangroves have been destroyed. Mangroves are a biological community that occurs in the intertidal zone along coastlines in tropical and subtropical areas. The major causes of the losses can be traced to human activities such as timber harvesting, the conversion of mangroves to aquaculture pools in Southeast Asia, and the clearing of forests for farming activities and housing developments.

FIGURE 3-9 • More than 100 countries have mangrove forests, but the most extensive ones are found in Bangladesh, Indonesia, Nigeria, Australia, Mexico, Malaysia, and Brazil.

Red Tide

A seasonal population explosion of toxic algae (dinoflagellates) that affects the ocean food chain is called the red tide. Red tides occur in blooms when water temperature, salinity, and nutrients reach levels that support a massive increase in the dinoflagellate population. This type of algal bloom can lead to high mortalities among marine organisms as toxins produced by the algae are released into the water. The large numbers of algae can also turn the water red, olive green, or yellow.

A concentration of 250,000 red tide organisms per liter of salt water is lethal to fish. As the algae die off, the decay of the organisms depletes the oxygen in the water, causing death to huge numbers of fish, shellfish, and crustaceans. The red tide was linked to the deaths of 39 manatees in Florida in the late 1970s. Both during and following a red tide, the beaches and shoreline may be covered with dead and dying fish, closing beaches for days or even weeks. Swimming in red tide water is not very harmful to humans, but doing so can cause throat and nose irritation and burning eyes. The poison released by the algae can also accumulate in the cells of shellfish such as clams and mussels, making them toxic to humans. Red tides are common on the northeast coast of the United States, in the Baltic Sea, and in the Adriatic Sea.

In 1999 scientists reported that pollution may not be the cause for the red tides. Their research reveals that it is the distribution of the microorganism *Alexandrium tamarensis* that causes the red tides. The organism, which appears in a dormant stage as a cyst in the ocean, becomes active as a swimming dinoflagellate under the right conditions. The dinoflagellates can wash into beds of mussels, oysters, and clams, where the shellfish ingest the microorganisms, making the shellfish poisonous to eat.

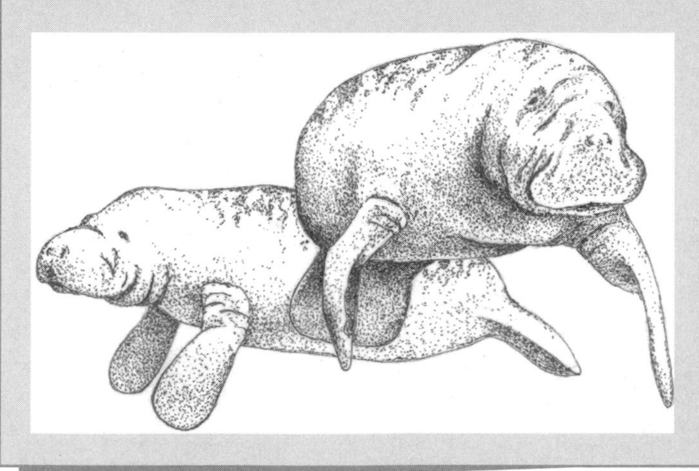

FIGURE 3-10 • Manatees are protected at the federal level by the Marine Mammal Protection Act of 1972 and the Endangered Species Act of 1973.

SALT MARSHES

Human activities are also harming salt marshes. Salt marshes are marine ecosystems characterized by fluctuating water levels, which are the result of tides, and the growth of salt-tolerant plants. One of the most important threats to salt marshes as well as to most other marine ecosystems is runoff from diverse land locations. These include bridges and roads (petroleum products from cars) and from farms and lawns (pesticides and fertilizers). Pollution disrupts the food web in the salt marsh by killing off some species and causing others to greatly increase in number.

Refer to Volume I for more information about salt marshes, mangroves, and coral reefs.

OIL POLLUTION

Oil pollution is any accidental or intentional discharge of petroleum into the environment from point sources such as pipelines, tankers, recreational watercraft, pleasure boats, and land and offshore drilling oil rigs. Other sources of oil pollution include spillage from recreational power boats as well as the runoff of gasoline and oil from the surface pavements of roads, parking garages, and gasoline stations. Most of this runoff finds its way into waterways, causing water pollution. Runoff from rivers and other waterways represents about 30 percent of all oil pollution sources in the ocean.

Large oil spills sometimes occur when tankers run aground or leak as a result of collisions. Such spills account for about 3 to 5 percent of

Presently, there are many double-hulled vessels in service of transporting oil. The double-hulled vessels provide more protection than single-hulled vessels in accidents at sea. (Courtesy of OSG Ship Management, Inc.)

TABLE 3-4 Time Capsule Major Oil Spills

Year	Spill/Effects
1967	***Torey Canyon* Oil Spill.** Torey Canyon runs aground off Cornwall, England, spilling about 175 tons of crude oil.
1969	**Santa Barbara Oil Well Blowout.** An oil well blowout near Santa Barbara, California, leaks about 2,700 tons of crude oil into the Pacific Ocean.
1973	***Corinthos–Edgar M. Queeny* Collision.** In a collision in the Delaware River in Marcus Hook. Pennsylvania, the port anchor of the *Edgar M. Queeny* penetrates the hull of the *Corinthos*, resulting in an explosion. The 20,000 tons of chemical cargo from the *Edgar M. Queeny* which includes gasoline, methanol, phenol, vinyl acetate, and styrene monomer, and the 272,000 barrels of crude oil carried on the *Corinthos* are released into the air as a result of the fire or into the waters of the Delaware River.
1976	***Argo Merchant* Oil Spill.** *Argo Merchant* runs aground off Nantucket, Massachusetts, releasing almost 25,000 tons of fuel oil into the Atlantic Ocean.
1976	***Hawaiian Patriot* Oil Spill.** *Hawaiian Patriot* catches fire, releasing almost 100,000 tons of oil into the Pacific Ocean.
1977	**Ekofisk Oil Well Blowout.** An oil well blowout results in nearly 27,000 tons of crude oil being spilled into the North sea.
1978	***Amoco Cadiz* Oil Spill.** *Amoco Cadiz* runs aground off Portsall, Brittany, spilling 226,000 tons of oil into the ocean.
1979	***Atlantic Empress–Aegean Captain* Oil Spill.** A collision between the *Atlantic Empress* and the *Aegean Captain* spills more than 370,000 tons of oil into the Carribbean.
1983	**Iran-Iraq War Oil Spill.** Iraq attacks wells of Nowuz oil field in Iran, resulting in 600,000 tons of oil being released into the Persian Gulf.
1988	**Ashland Oil Spill.** Collapse of a storage tank releases nearly 2,500 tons of oil into the Monangahelia River near Pittsburgh, Pennsylvania.
1989	***Exxon Valdez* Oil Spill.** The *Exxon Valdez* runs aground in Prince William Sound off the Alaskan coast, releasing about 37,000 tons of oil into the sound.
1989	***Kharg 5* Oil Spill.** The *Kharg 5* oil tanker catches fire off the Canary Islands, releasing 75,000 tons of oil into the surrounding water.
1991	**Persian Gulf War Oil Spills.** Iraqi troops deliberately spill oil stored at Sea Island Terminal in Kuwait and set the oil fields ablaze. Much of the oil enters the water of the Persian Gulf or spreads over the surrounding land. That which does not seep into water or soil is released into the air through the fires that burn for well over a year.
1994	**Komi Republic Oil Spill.** A dike constructed to contain oil leaking from a pipeline near Usinsk in northern Russia (just below the Arctic Circle) collapses, releasing nearly 102,000 tons of oil onto the Siberian tundra.
1999	***New Carissa* Oil Spill.** The *New Carissa* runs aground in Coos Bay off the coast of Oregon. The tanker carries more than 1,500 tons of fuel oil, some of which leaks into the water over a period of days. Some remaining oil is pumped from the tanker to containers on shore. Next, the tanker is set ablaze, believing release of hydrocarbons into the air will pose less of a threat than its release into the water. Finally, the tanker is towed into the ocean, where it is expected to rest 1,825 meters (6,000 feet) below the surface. Temperatures on the ocean floor are believed to be cold enough to keep the oil in a solid state that will pose little environmental threat.

Source: John Mongillo and Linda Zierdt-Warshaw, *Encyclopedia of Environmental Science* (Phoenix, Ariz.: Oryx Press, 2000).

TABLE 3-5 Sources of Oil Pollution in Oceans

Source	Percent
Runoff from rivers and waterways	31 percent
Tanker activity: loading, unloading, etc.	20 percent
Sewage plants and refineries	13 percent
Underwater seepage, cracks in ocean floor	9 percent
Pleasure boats, fishing vessels, ferries, etc.	9 percent
Oil tanker accidents	3 percent
Other	15 percent

ocean oil pollution. However, a single large spill can devastate an area for a long time. The *Exxon Valdez* oil spill, for example, which occurred in the Prince William Sound off the southern coast of Alaska, released about 37,000 tons of oil into the sound and caused extreme damage to ocean ecology in that area.

Oil pollution is a particular problem in the world's oceans, where it can have devastating effects on wildlife and ecosystems. Oil is toxic and directly kills small animals, such as fish, shrimp, crabs, and other shellfish. Other animals are poisoned indirectly when they feed upon these oil-soaked prey. Birds, sea otters, and other large animals are harmed by oil when it forms a slimy coating over their feathers or fur. This thick, unrefined black liquid does not wash off easily in water and destroys the insulating properties of feathers and fur. As a result, oil-contaminated animals develop *hypothermia* and often freeze to death.

Most oil pollution in the ocean results from less spectacular events. These include accidental and illegal discharges of oil when oil tankers are loaded and unloaded and minor spillages from tankers as they transport oil across the ocean.

LAWS TO PROTECT THE OCEAN

Today laws help protect against oil pollution in the oceans. For example, the Oil Pollution Act of 1990, a federal law of the United States, increases the liability of oil tanker owners by having them abide by strict regulations regarding oil transportation.

Ocean Dumping Act

Until the 1970s, the world's oceans were largely considered by industry and certain governments as convenient out-of-site, out-of-mind dumping grounds. Since then, however, public and political perceptions have changed about ocean dumping. People and governments worldwide now recognize that potential polluters should deal with their own wastes rather than dump them. Today there are a number of laws making it illegal to dump wastes into the ocean.

In the United States, the most significant law protecting against ocean dumping is the Ocean Dumping Act. Passed in 1972, this important law bans ocean dumping of chemical and biological warfare agents and of high-level radioactive waste. Amendments in 1988 extended this ban to sewage sludge, industrial wastes, and medical wastes. The Ocean Dumping Act also authorizes research on the effects of ocean pollution, overfishing, oil spills, and other human-induced problems. Provisions added in 1992 established a national coastal water–monitoring program to evaluate the health and quality of ocean waters and the pollution sources that affect them.

On an international level, the London Convention organization protects ocean waters worldwide. All the countries in the organization promote the effective control of all sources of pollution in the marine environment. They have initiated steps to prevent the dumping of waste and other matter that is hazardous to human health and marine life. The London Convention has now been signed by more than 70 countries, including the United States, Canada, Japan, France, Germany, and China.

Vocabulary

Alkaline A substance that can neutralize an acid.

Aquifers Any rock formations containing water.

Chemoautotrophs Organisms that synthesize nutrients from inorganic chemicals.

Dissolved oxygen Oxygen molecules dissolved in water; needed by organisms that live in water.

Fecal coliform A bacteria that lives in the colon of warm-blooded mammals such as humans.

Heavy metals Metallic elements such as lead, mercury, and arsenic. Excessive concentration of heavy metals in the human body can pose a health threat.

Hypothermia A subnormal body temperature.

Pathogens Microorganisms and viruses that cause diseases.

Watershed An area into which all water drains.

Activities for Students

1. What are the biological and chemical processes that cause a person to be sick from eating the meat from an animal that drank water contaminated by coliform bacteria?
2. Given that the world's supply of freshwater compares to about a tablespoon of water from a gallon of water, what ways can saltwater be treated to become a source of freshwater? What would be the resulting extra sediments or nutrients left from these treatments?
3. Visit your local water treatment plant that services the water supply to your area. What besides H_20 is in the water? Why?
4. Contact your local PADI dive shop or the editors of *Scuba Diver* magazine to ask what professional divers and dive companies are doing to help preserve the coral reefs around the world.

Books and Other Reading Materials

Air and Waste Management Association. *Pollution Prevention for Our Land, Water and Air.* Pittsburgh, Penn.: Air and Waste Management Association, 1993.

Bolling, David M., ed. *How to Save a River: A Handbook for Citizen Action.* Sponsored by the River Network. Washington, D.C.: Island Press,1994.

Gorman, Martha. *Environmental Hazards: Marine Pollution: A Reference Handbook.* Contemporary World Issues Series. New York: ABC-CLIO. 1993.

Jorgensen, Eric P., ed. *The Poisoned Well: New Strategies for Groundwater Protection.* Sponsored by Sierra Club Legal Defense Fund. Washington, D.C.: Island Press, 1989.

Nonpoint Pointers: Managing Wetlands to Control Nonpoint Source Pollution. Pointer No. 11 Order No. EPA841F96004K. National Service Center for Environmental Publications, P.O. Box 42419, Cincinnati, OH 45242-2419.

Schneider, Paul. "Clear Progress: Twenty-five Years of the Clean Water Act." *Audubon 99* (September–October 1997): 36–47, 106–7.

Solo-Gabriele, H., and S. Neumeister. "U.S. Outbreaks of Cryptosporidiosis." *Journal of the American Water Works Association*, September 1996.

Talen, Maria. *Ocean Pollution.* Lucent Overview Series. San Diego, Calif.: Greenhaven Press, 1991.

U.S. Environmental Protection Agency. Office of Water, Office of Pesticides and Toxic Substances. *National Survey of Pesticides in Drinking Water Wells.* Publication 570/9-90-015. Washington, D.C.: GOP, 1990.

Websites

Environmental Protection Agency, Summaries of Major Environmental Laws, http://www.epa.gov/region5/defs/index.html

Environmental Protection Agency Oils Spill Program, http://www.epa.gov/oilspill/overview.htm.

Environmental Protection Agency's Office of Wetlands, Oceans, Watersheds for Nonpoint Source Information, http://www.epa.gov/owow/wetlands/wetland2.html and http://www.epa.gov/swerosps/ej/

The Groundwater Foundation, http://www.groundwater.org

National Groundwater Association Home Page, http://www.h2o-ngwa.org

Nonpoint Source Pollution Control Program, http://www.epa.gov/OWOW/NPS/whatudo.html and http://www.epa.gov/OWOW/NPS/

Ocean Planet, http://www.seawifs.gsfc.nasa

Sierra Club, "Happy Twenty-Fifth Birthday, Clean Water Act," http://www.sierraclub.org/wetlands/cwabday.html

U.S. Geological Survey, *Water Resources of the United States*, http://www.water.usgs.gov

CHAPTER 4

Solid Wastes: Garbage, Trash, and Litter

In 2000, the U.S. Environmental Protection Agency (EPA) estimated that approximately 232 million tons of municipal solid waste in the United States was generated by household and industrial sources. This number represents about two kilograms (4.5 pounds) of waste per person per day up from a little over one kilogram (2.7 pounds) per person per day in 1960. About 55 percent of the solid waste is disposed of in landfills, 31 percent is reused or recycled, and 14 percent is burned at incinerator plants.

This chapter concentrates on solid wastes—those materials that are handled and discarded as solids such as garbage, trash, and litter. Hazardous and *toxic* wastes are solid, liquid, or gaseous wastes that have characteristics or contents that makes them potentially hazardous.

Refer to Chapter 5 for more information about hazardous and toxic wastes.

SOLID WASTES

Garbage

A major solid waste problem for many urban areas in the world is the removal of garbage. According to one estimate, various countries produce about 720,000 million tons of garbage per year. The United States produces the most garbage. Other leading garbage-producing countries include Australia, Canada, and Switzerland.

The EPA describes garbage as food waste from animals or plants. Garbage results from the handling, storage, packaging, sale, preparation, cooking, and serving of foods. Garbage is disposed of in landfills or burned at incinerator facilities. In some countries, however, garbage is dumped in open pits where food wastes are exposed to insects and rodents. Each year, millions of people become sick or die from such *unsanitary* disposal of garbage.

Trash and Litter

Trash is solid household and business waste materials. Trash includes yard wastes, paper products and cardboard goods, clothing, glass, plastic, appliances, wood products, and aluminum containers. Stereos, telephones, and computers are a small but rapidly growing portion of discarded trash.

DID YOU KNOW?

According to a garbologist, paper wastes take up three to four times the volume of plastic wastes in a landfill.

Perhaps one of the most essential services in any inner-city neighborhood is the removal of trash. As a result of the high proportion and constant turnover of rental units, the trash collection and removal of bulk items such as old sofas, refrigerators, and other discarded furniture has increased. Trash is usually picked up by vehicles and brought to local sanitary landfills. Much of it including paper, aluminum, and glass containers, can be recycled.

In the United States, some states export their trash to other states. Pennsylvania is one of the leading states in importing trash. In 1998, the state imported about 6 million metric tons of trash. Other importers of trash include Virginia, Indiana, Michigan, and Illinois. The U.S. Congress wants to limit the interstate transportation of trash.

Materials scattered around in disorder, especially rubbish, is litter. According to Keep America Beautiful research, there are specific primary sources of litter:

- Household trash handling and its placement at the roadside curb for collection
- Dumpsters used by businesses
- Loading docks, construction, and demolition sites
- Illicit roadside dumping
- Trucks with uncovered loads
- Pedestrians and motorists

One of the biggest problems with litter is that it stays around a long time. Much litter is not *biodegradable*. For example, uncollected foil products, plastic-coated wrappers, and metal containers can litter an area, such as a roadside, from several weeks to several years. Street litter is often swept away with rainwater into storm drains. From there,

TABLE 4-1 National Average of Trash Breakdown

38%	Paper
18%	Yard Trimmings
8%	Metals
8%	Plastic
7%	Glass
7%	Food Waste
14%	Other

Solid wastes include various waste products. About 55% of all solid waste is disposed of in landfills.

it can flow into the ocean and wash up on the beaches. Organizations such as Keep America Beautiful have been formed to combat the litter problem. Many cities have written laws establishing fines for littering.

SOLID WASTE MANAGEMENT FACILITIES

A waste management facility is specially designed, constructed, and operated to manage various types of wastes. Examples of solid waste management facilities include sanitary landfills, recycling stations, transfer stations, and incineration or combustion facilities.

Landfills and Sanitary Landfills

Throughout much of our history, cities and towns used open landfills, called dumps, to dispose of trash and garbage. The dumps were often constructed in lowland areas that had little value for farming or housing. Some of the dumps were located in areas such as wetlands and marshes or in old and abandoned mines. Presently, in much of the developed countries, the open landfills and garbage dumps have been replaced by sanitary landfills as disposal sites for solid wastes.

In the United States, about 55 percent of all solid waste is buried in sanitary landfills. In general, a sanitary landfill is a disposal site for deposits of nonhazardous refuse such as solid waste, trash, rubbish, and garbage. However, there is a certain amount of household hazardous materials that is also collected and buried in landfills. These household items, such as paint, cleaners, oils, batteries, and pesticides contain hazardous components. If mishandled, they have the potential to be dangerous to human health and the environment.

Much of the waste that ends up in the landfill includes paper products, yard trimmings, food scraps, plastics, bottles, and metals. Paper products are the major waste material in landfills. On average, paper wastes account for more than 37 percent of a landfill's contents. This proportion of paper wastes has held steady for decades, and in some places discarded paper products in landfills have actually risen. Newspapers alone can take up as much as 13 percent of the space in U.S. landfills.

Landfills can be a source of pollutants that can escape into the environment if not monitored. *Organic material* is dumped into landfills every day. The organic material gradually degrades, naturally producing various byproducts. The liquid produced by the degrading waste and by water filtering through the landfill is called leachate. The leachate is made up of a highly polluted mixture of water and other products from the buried wastes. The pollutants in leachate can include organic substances and *inorganic material* that can contaminate groundwater or surface water. In addition, biological contaminants such as bacteria can be present in the leachate.

Landfills can also be a major source of air pollutant gases. These gases are emitted into the atmosphere by the decomposing garbage and other wastes. The gases include carbon dioxide, hydrogen sulfide, and methane. Some of the gases can be hazardous if they are not collected or vented through a pipe into the atmosphere. As an example, methane may spread out from a landfill to nearby homes or underground structures, such as sewers, where explosions may occur. Methane also contributes to the formation of smog conditions. Landfill methane gas can also be collected and used as a fuel. Hydrogen sulfide is a foul smelling and poisonous gas that can accumulate in the buried waste and can pose a hazard to nearby sewers or basements. To monitor and control landfill pollution, the sites have to be designed and well built to protect people and the environment.

LANDFILL DESIGN

The sanitary landfill is built primarily to reduce *unsanitary* conditions to human health. The refuse brought to the landfill is spread out over an area and then flattened down by heavy machinery, tightly compressing the material. The compacting is done to reduce the volume of the refuse. At the end of each day, the crushed refuse is covered by soil or other material. Doing this on a daily basis helps reduce odors, wind-blown debris, and other nuisances. Some large landfills can handle more than 3,500 metric tons of garbage per day. The buried material is graded into a mound that slopes at about a 30-degree angle, which helps keep the material stable. Over time, landfills grow in size; some develop into great mounds of earth.

To prevent the buildup of leachate, sanitary landfills are lined on the bottom with dense clay (clay liner). They are then sealed with layers of plastic, which acts as a barrier to prevent leachate from entering groundwater, and eventually are capped with low-permeability materials. The daily spreading of soil over the landfill also prevents rain from leaching the dumped wastes. But heavy rainfall can cause serious erosion of landfills and polluted runoff. Thus, landfills include runoff control features. Often a leachate collection system is built around a landfill to capture leachate before it escapes into the environment. Any leachates collected at the bottom of the landfill are pumped out, collected, and then treated to remove contaminants. The left-over sludgelike material can be burned or used as fertilizer. If the sludge contains *toxic* materials, it can be transported to a hazardous waste site.

To prevent methane and other gases from escaping from modern landfills, a collection system of pipes is built into the landfill site. The pipes allow gases to be vented into the atmosphere, incinerated, treated, or collected in storage areas and used as fuel. In 1996, the EPA required the largest landfill operations, about 280 sites, to reduce their air pollutant emissions by 90 percent. In this case, landfill operators had to drill wells into the sites to collect the gases before they were released into the atmosphere.

DID YOU KNOW?

In Wayne, Michigan, methane gas from a landfill is processed and used to produce electricity for three local motor vehicle factories.

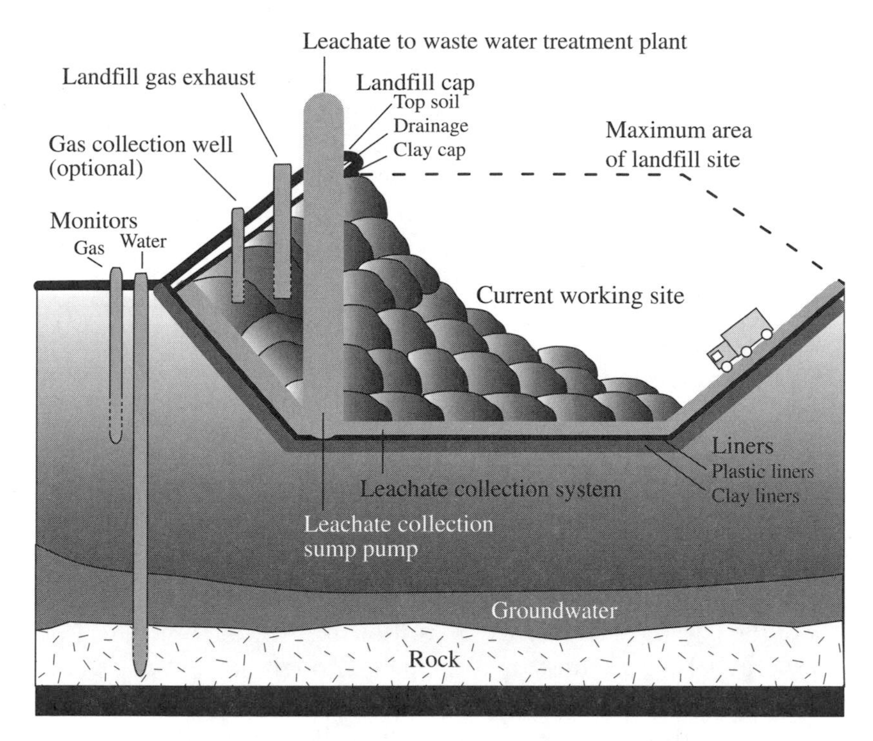

FIGURE 4-1 • Diagram of a Typical Sanitary Landfill The sanitary landfill is built to reduce the nuisance of solid waste and the threat of unsanitary conditions to human health. Methane gas from some landfills can be used as a fuel to produce heat and electricity for local buildings.

Before a landfill is closed, it must be sealed with landfill caps. Landfill caps are specially designed covers that are constructed to seal the tops of old or new landfills. The cap reduces leachate production by preventing water from entering the landfill; it also reduces erosion and helps control landfill gases. Sometimes the finished landfill is purchased and used as a recreational site or golf course.

The number of active landfills in the United States is steadily decreasing—from 8,000 in 1988 to 1,967 in 2000. The standards and regulations for landfills are monitored and supervised by state governments and may be operated by private businesses or by counties, cities, or towns. The EPA regulates only those municipal landfills that service more than 100,000 people.

One of the world's largest landfills was the Fresh Kills landfill located on Staten Island in New York. The Fresh Kills landfill had been operating since 1948. Each day, garbage trucks dumped about 12,000 tons of wastes into the landfill. In time, the landfill reached the height of a 17-story building spread over 1,200 hectares (3,000 acres). The landfill was closed in March 2001, and now the garbage is hauled to nearby states. Whether or not these states will continue to accept New York garbage for very long is open to debate.

Incineration

Approximately 14 percent of all the trash in the United States is burned in incinerators. In 2000, there were in operation 102 incinerators with the capacity to burn nearly 96,000 tons of municipal solid waste per day. Incineration is a burning process using controlled high temperatures to reduce waste materials to noncombustible ash, carbon dioxide, and water. Incineration is used for two types of wastes—municipal solid waste and hazardous waste. (The incineration of toxic and hazardous materials is covered in Chapter 5.)

Most municipal solid waste incinerators burn wastes such as garbage, paper, and cardboard after rubber, metal, and glass wastes are removed. These incinerators can reach about 500°C (915°F). Some of these incinerators that process large amounts of solid waste per day can be used to produce steam to be used for local heating. The steam can also be piped to electric power plants to produce electricity.

The advantages of incineration are that it reduces the volume of original waste by 50 to 75 percent and vaporizes the liquid wastes. The remaining ash is periodically removed, transported, and deposited into sanitary landfills. The disadvantages of incineration include the emission of toxic ash, sulfur, and nitrogen oxides from chimney stacks into the atmosphere that contributes to air pollution and acid rain. As a result of the Clean Air Act (CAA) legislation, many modern incinerators have scrubbers or electrostatic precipitators to remove and trap pollutants before they are emitted into the atmosphere. Another environmental concern is that the ash residue, usually disposed of in landfills, can contain hazardous materials such as heavy metals that can leak into the groundwater.

In waste-to-energy plants, tons of garbage are hauled in each day to produce electricity. (Courtesy of National Renewable Energy Laboratory)

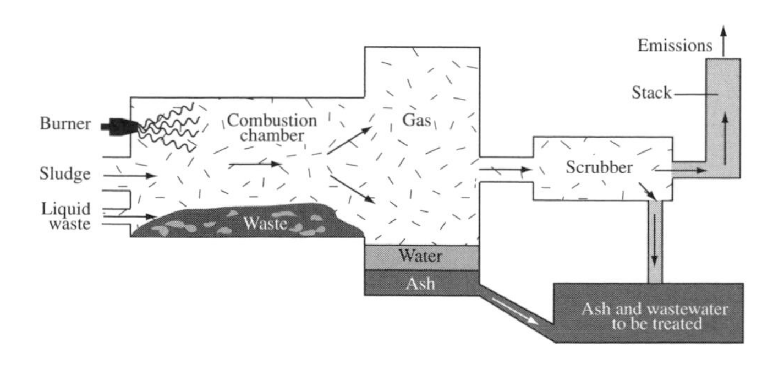

FIGURE 4-2 • Solid, liquid, and sludge wastes are burned in the combustion chamber. The byproducts such as wastewater and ashes are collected and treated before disposal.

Source Reduction and Recycling

Environmental groups emphasize that source reduction and recycling are better alternatives to waste reduction than by incineration. Source reduction and recycling prevent or divert materials that would normally be thrown away in a landfill or burned in an incinerator.

Sources Reduction

Source reduction involves the altering of the processes or practices to reduce the amount of waste generated at its source or place. An example of source reduction would be a toy company redesigning the packaging of its toys into smaller boxes. Another example would be an automobile company identifying all the plastic components it uses in making its cars. When these cars were no longer operative, the different plastics in the discarded cars would be collected, sorted, and recycled. Other practices of source reduction such as grass cycling, backyard composting, and the two-sided copying of paper have yielded substantial benefits through source reduction. Source reduction has many environmental benefits. It prevents emissions of many greenhouse gases, reduces pollutants, saves energy, conserves resources, and cuts down on the need for new landfills and incinerators.

Recycling

About 31 percent of all trash in the United States is recycled. Recycling is also widely practiced in Japan and Europe. In fact, Japan recycles 50 percent of its garbage. The Netherlands, Germany, Austria, Switzerland, and Finland are leading countries in recycling of paper goods.

Recycling is a method for conserving natural resources that involves collecting discarded materials and reprocessing these materials to make new products. A common example of recycling is the collecting of discarded aluminum beverage cans for reprocessing into new aluminum products. Recycling natural resources provides several benefits to the environment. First, the recycling of materials reduces the need to use raw materials, extending the useful life of reserves of such resources. This benefit is particularly important to the conservation of *nonrenewable resources*.

Recycling reduces problems associated with waste disposal by decreasing the volume of materials that must be deposited into landfills or disposed of in some other way, such as incineration. Like nonrenewable resources, land available for landfills or other means of waste disposal is in limited supply. Thus, finding ways to reduce the overall amount of materials that need to be discarded can help to conserve land space. Recycling also reduces the potential for discarded materials to give off toxic or otherwise harmful substances to the environment as they break down.

Another benefit of recycling is that it often helps to conserve energy. It has been estimated, for example, that reprocessing beverage cans to reclaim the aluminum they contain uses as much as 95 percent

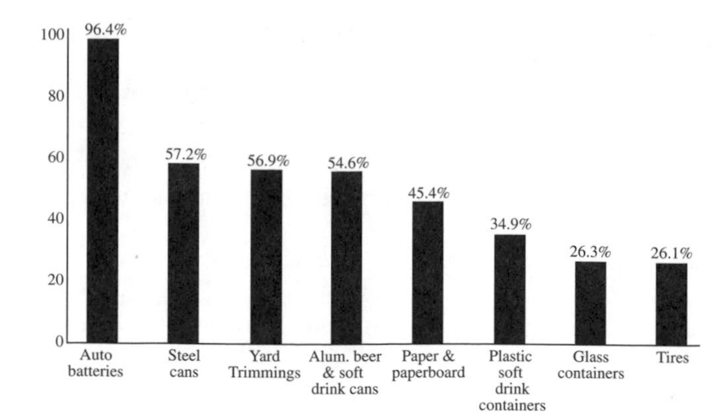

FIGURE 4-3 • Recycling Rates of Selected Materials *Source:* Environmental Protection Agency.

DID YOU KNOW?

New Jersey, New York, Maine, South Carolina, and South Dakota rank very high in recycling. How does your state rank?

less energy than obtaining aluminum from ore. This energy savings helps to conserve the vital fossil fuels (coal, petroleum, and natural gas) that often are burned to process ores. The burning of such fuels is a major source of air pollution, for when burned, they release such substances as carbon dioxide, sulfur dioxide, smoke, and ash into the atmosphere. Burning less of these fuels helps to reduce the amounts of these emissions that are released into the atmosphere while also helping to conserve vital reserves of these fuels.

In the last quarter of the twentieth century, many nations have initiated recycling campaigns to help conserve natural resources. Among the most common recycled materials are paper, aluminum, glass, plastic, steel, motor oil, and many kinds of battery systems. More recently, methods for recycling rubber, lead, zinc, copper, and brass have also been implemented. Once recycled, these materials are used to make a variety of products, including many of the component parts used in the manufacture of new automobiles, resurfacing materials for roadways, and reprocessed versions of the original products (glass containers, cans, and newspapers) from which the recycled materials were obtained.

BOTTLE BILL

A bottle bill is a container-deposit legislation that helps reduce litter and encourages recycling. Such laws require that a small deposit be collected when consumers purchase a beverage container. The consumers receive a refund when the container is returned. Store owners or special outlets are responsible for collecting the containers and storing them for collection or recycling.

Oregon passed the first bottle-recycling bill in 1972. Oregon's Beverage Container Act (Bottle Bill) is a 30-year success story. Within four years after implementation of the bill, 90 percent of Oregonians favored the law, and it continues to be one of the most popular pieces of legislation ever passed in the state. When it was passed, the bill was viewed primarily as a litter control measure. After the Bottle Bill's first year, nearly 4 million fewer beverage containers had been thrown away than the preceding year. Within two years, litter from beverage containers had

TABLE 4-2 Products Made from Recycled Waste

Waste Material	Product
Paper	printing paper, writing paper, packaging, construction paper, hardboard
Sludge	compost, roadbeds
Rubber (tires)	pavement, retreads
Plastic	pipes
Glass	ceramic bricks, concrete
Iron and steel	cast-iron pipes, structural shapes
Aluminum	siding
Slag	cement
Fly and bottom ash	roadfill, asphalt, cement
Sulfur	asphalt cement
Oil	oil
Refuse-derived fuel	energy
Various wastes	automobiles, hand tools
Various chemicals	paint, soap and wax
Wood, metal, textiles	office furniture
Kiln, lime, and gypsum dust	fertilizer

Source: U.S. General Accounting Office, 1980.

TABLE 4-3 Types of Batteries Accepted by INMETCO Recycling Facility

Battery Systems	Applications
Lithium Ion	computers, portable phones
Lead Acid, Automobile	automobiles, tractors, marine
Magnesium	transmitters, aircraft transmitters
Mercury	hearing aids, military sensors, and detection tools
Nickel Cadmium, Industrial	railroads, communications, aircraft, utilities
Nickel Cadmium, Dry-sealed	portable phones, tools, computers, appliances
Nickel Iron, Industrial	railroad signals
Nickel Metal Hydride	portable phones, computers
Zinc Carbon	flashlights, toys, radios, instruments

INMETCO, a company located in Pennsylvania, recycles the nickel, chrome, and iron in the byproducts of stainless steel production so that the materials can be reclaimed and reused instead of landfilled. The company also recycles a large variety of batteries.

dropped by 83 percent. Within 15 years, beverage containers accounted for only 4 percent of roadside litter, down from 40 percent prior to the Bottle Bill. Most impressively, the bill created a broader anti-litter ethic. Within two years, Oregon's roadside litter was cut almost in half.

Curitiba, Brazil, Deals with Garbage Disposal

Several cities are initiating programs to deal with the garbage problem. One unique campaign, called the Garbage Purchase Program, takes place in Curitiba, Brazil, a city with a population of about 2 million people. Because of narrow streets in certain sections of the city, garbage trucks cannot drive along routes to many of the homes. The program is therefore designed to have residents fill up their garbage bags and bring them to the trucks waiting along outlying streets. In exchange for the garbage, the residents receive compensation in the form of surplus food, bus tokens, and school supplies. City officials believe that this program does not cost any more than if they hired garbage collectors to go door-to-door in the city. The program is also well received by the city dwellers. More than 35,000 families participate in this program. Other city programs make use of source separation, recycling, and composting to reduce the garbage being transported to sanitary landfills.

Massachusetts, Maine, Delaware, Michigan, Connecticut, Vermont, Iowa, New York, and California are other states that have bottle laws. Those states in favor claim that bottle bill legislature not only reduces litter because containers can be refilled several more times but also conserves energy by producing fewer new containers. Those states opposed claim that bottle bill laws increase costs for labor to handle the returnable containers, create sanitary problems in handling the returnables, and raise beverage costs for the consumer. Opponents also contend that many kinds of containers cannot be recycled or refilled. However, public support for bottle-deposit laws is high and the U.S. bottle bill law has not been repealed. Outside the United States, there are container deposit programs in Sweden, Denmark, Norway, and the Netherlands. Several provinces in Canada also have bottle-deposit laws.

There are several key benefits to recycling:

- Reducing the need for landfilling and incineration
- Saving energy and preventing pollution caused by the extraction, processing, and manufacturing of materials used to make products
- Decreasing emissions of greenhouse gases that contribute to global climate change
- Conserving natural resources such as timber, water, and minerals
- Helping sustain the environment for future generations

DID YOU KNOW?

Some recycled plastics can be melted down and spun into clothing fibers. It takes about 25 one-liter (33.8 oz) plastic bottles to weave one jacket.

COMPOSTING

Composting is the process of recycling various organic wastes. Composting uses the process of *aerobic decomposition* of organic matter by microorganisms. These organisms include bacteria, fungi, and invertebrates.

The composting process is completed when the recycled organic materials are transformed into a dark humus called compost that can be used as a natural fertilizer.

Composting can take place in backyards and in large institutional and municipal facilities. Scientists believe that composting is an effective and inexpensive option in reducing the volume of organic materials sent to landfills for disposal.

For many years, solid waste–composting plants have been operating in western Europe, Israel, and Japan. In the United States, yard waste composting is a key tool in addressing the municipal solid waste stream because yard waste accounts for nearly one-fifth (29.8 million tons) of the municipal solid waste generated. Some communities have begun to conduct large-scale centralized composting of yard waste in an effort to save landfill capacity. Individuals are also helping by composting yard waste in their backyards, rather than by bagging grass clippings or other yard wastes for municipal sanitation pickup. These activities are actually classified as source reduction. Composting of yard waste has seen tremendous growth in recent years. In 1985 in the United States, the amount of yard waste recovered was negligible, less than 50,000 tons. By 1995, the amount of yard waste recovered had increased to 9 million tons.

In large-scale composting, municipalities collect yard waste and take it to a central location. At the site, the material is piled into triangular-shaped rows, between one meter (three feet) and three meters (nine feet) high. About once a month, heavy equipment is used to mix the piles for aeration. The temperature and moisture of the piles are checked twice a week. The finished compost is then sold, given away, or used in public works projects.

High school teacher records daily measurements of a compost pit. To get active composition completed in two to six weeks in the compost, the process requires three key ingredients: aeration, adequate moisture, and proper temperature. (Courtesy of Nancy Trautmann)

What Goes on in the Compost Pile?

The process of composting can take place quickly or slowly depending on which methods are used. To complete active composition in two to six weeks requires three key ingredients: aeration, adequate moisture, and proper temperature.

Oxygen, through the process of aeration, is needed by the microorganisms or invertebrates to successfully decompose the materials in the compost pile. Aeration provides the oxygen that would otherwise be lacking in the pile. Turning the compost pile is an effective means of adding sufficient oxygen for efficient decomposition.

The moisture content of the material in the pile should be between 40 and 60 percent. Compost experts use the squeeze test. They squeeze a handful of material and compare the moisture content to the ideal level: that of a well-wrung sponge. If the moisture content is below 40 percent, the activity of the microorganisms will slow down. If the pile is too wet, above 60 percent, aeration is hindered, decomposition slows down, leaching occurs, and there is a definite odor from anaerobic decomposition. A wet pile can be corrected by adding more dry material.

Microorganisms generate heat as they decompose organic materials. In the temperate zone, an efficient compost pile will generate temperatures between 32° C and 60° C (90° F to 140° F). Temperature that is too high will inhibit the activity of the organisms. Cool temperatures will advance the decomposition process, but at a slower rate.

Another factor in the composting process is the ratio of carbon to nitrogen, which are two important elements in composting. The bulk of the organic material should be mostly carbon, with just enough nitrogen to help in the decomposition. The ratio should be 30 parts carbon to 1 part nitrogen (30:1). Carbon is considered to be the "food," and nitrogen to be the "digestive enzyme." Leaves are a good source of carbon, whereas grass and manures are sources of nitrogen.

At Michigan State University, researchers believe composting may have the potential of helping to clean up industrial waste sites. Composting wastes is a cleaner and less expensive solution than using an incinerator or having to provide offsite disposal. Composting may not work on extremely polluted soil, but it could be used to clean up land contaminated with polychlorinated biphenyls (PCBs). Polychlorinated biphenyls were used widely as industrial lubricants for 50 years until scientists linked them to cancer, reproductive disorders, problems with brain development, and other diseases. The first step in the experiment is to see whether organic material alone can detoxify PCBs in a timely manner.

REGULATING SOLID WASTES

Resource Conservation and Recovery Act

In 1976, the Resource Conservation and Recovery Act (RCRA) established guidelines for disposing solid wastes. The legislation includes four main goals:

- Protecting human health and the environment from possible hazards involving waste disposal
- Reducing the total amount of generated waste material

- Conserving energy and natural resources
- Ensuring environmentally sound waste management practices

Vocabulary

Aerobic decomposition The biodegradation of materials by aerobic microorganisms that require oxygen to live.

Biodegradable Susceptible to the metabolic breakdown of materials into simpler organisms by living organisms.

Inorganic material Chemicals or substances that do not contain carbon atoms.

Nonrenewable resources Resources such as coal and petroleum that are not replaceable.

Organic material All living matter that contains carbon atoms.

Toxic Poisonous or unhealthy.

Unsanitary Not healthy, not clean.

Activities for Students

1. Research the disposal of garbage during the Middle Ages. Compare your findings with our current system.
2. Begin or support a recycling program at your school. Include an education component.
3. Contact your local waste management and recycling agencies. Learn about the ways in which they handle waste and recyclable materials.

Books and Other Reading Materials

Berthoid-Bond, Annie. *Clean and Green: The Complete Guide to Nontoxic and Environmentally Safe Housekeeping.* Woodstock, N.Y.: Ceres Press, 1990.

Brandt, Anthony. "Not in My Backyard." *Audubon* 99 (September–October 1997): 58–62, 86–87, 102.

Environmental Industry Associations (EIA), 4301 Connecticut Avenue NW, Suite 300, Washington, DC 20008. The association publishes a report, *Landfill Capacity in North America.*

Saign, Geoffrey C. *Green Essentials: What You Need to Know about the Environment.* San Francisco: Mercury House, 1994.

Seymour, John, and Herbert Girardet. *Blueprint for a Green Planet: Your Practical Guide to Restoring the World's Environment.* New York: Prentice Hall, 1987.

U.S. Environmental Protection Agency. *RCRA: Reducing Risk from Waste.* United States Environmental Protection Agency, Solid Waste and Emergency Response (Pub. 5305W). Washington, D.C.: GPO, 1997.

U.S. Environmental Protection Agency. *Solid Waste and Emergency Response.* Pub. 5305W. Washington, D.C.: GPO, 1997.

Websites

BioCycle, Journal of Composting & Organics Recycling, http://www.jgpress.com

Composting News, http://www.recycle.cc

Cornell Composting, http://www.cfe.cornell.edu/compost/Composting_Homepage.html

Environmental Protection Agency Office of Solid Waste and Emergency Response—Composting, http:www.epa.gov/epaoswer/non-hw/compost/index.htm

Keep America Beautiful, http://www.kab.org

Landfills: An Issue Confronting Our Sustainable Use of the Land, http://www.lalc.k12.ca.us/uclasp/issues/landfills/landfills.htm

CHAPTER 5

Hazardous and Radioactive Wastes

The United States generates millions of tons of hazardous wastes per year. These wastes come from businesses, factories, nuclear power plants, farms, hospitals, petroleum refineries, chemical manufacturers, and other industries. As mentioned in Chapter 4, your home is also another source of hazardous wastes. Some of these wastes include such common items as paint, cleaners, oils, batteries, insect sprays, and medicines. As much as 45 kilograms (100 pounds) of household hazardous wastes are found in an average American home. If mishandled or spilled, these products, can be dangerous to your health and family as well as the environment.

HAZARDOUS WASTES

A hazardous waste is one that is poisonous (toxic), corrosive, or explosive, or it can catch fire (ignitable). Hazardous materials (HAZMAT) include solid, liquid, gaseous, organic, and inorganic substances. When these wastes are discarded, abandoned, or disposed of, they pose a potential risk to the health of humans, other organisms, or the environment. The U.S. Environmental Protection Agency (EPA) has developed a list of over 500 specific hazardous wastes.

Toxic Wastes

A toxic waste is a hazardous solid, gas, or liquid waste that is poisonous to human health when ingested or inhaled. The EPA and the U.S. Department of Health and Human Services (DHHS) have identified the major toxic wastes. Some of the major ones include arsenic, lead, mercury, benzene, polychlorinated biphenyls (PCBs), cadmium, chromium, and perchloroethylene.

ARSENIC

Arsenic (As) is a gray, heavy metal. It forms poisonous compounds used in making of pesticides or can be released into the environment as a byproduct from the smelting of copper and lead. Most naturally occurring arsenic compounds have no smell or taste, even when present in drinking water.

TABLE 5-1 Table Properties of Hazardous Materials

corrosive	capable of causing burns to the skin, eyes, or other body surfaces. Strong acids, such as nitric acid, and sulfuric acid are corrosive. Strong bases such as sodium hydroxide are corrosive.
ignitable	capable of easily catching on fire. Propane is an ignitable gas.
toxic	if eaten, breathed, or absorbed through the skin, is capable of causing harmful short-term or long-term health effects. Examples of toxic substances include lead and many pesticides.
oxidizing	capable of producing oxygen, which could cause a fire or explosion. Hydrogen peroxide is a strong oxidizing chemical, known as an oxidizer.
reactive	capable of reacting to produce poisonous gases. Compounds containing cyanide or sulfide may be reactive hazardous substances.
etiological	containing biological matter, such as bacteria or viruses, that may cause disease.

High levels of inorganic arsenic can cause thickening and discoloration of the skin. Other health problems may include stomach pain, nausea, vomiting and diarrhea, as well as numbness in the hands and feet. Arsenic is a *carcinogen* and can cause death in extreme doses.

LEAD

Lead (Pb) is a soft, heavy, tasteless, naturally occurring heavy metal found in small amounts in all parts of the environment, including the air, food, water, and soil. Compounds of lead are hazardous substances and air pollutants. Today lead is used in the manufacturing of gasoline, paints, plumbing supplies, roofing supplies, pesticides, and batteries. Lead is also used in radiation shields for protection against X-rays and in surgical equipment and computer circuit boards.

In adults, high levels of lead exposure may decrease reaction time and possibly affect the memory. Such lead exposure may also cause weakness in the fingers, wrists, or ankles and may increase blood pressure. At even higher levels of exposure, lead can severely damage the brain and kidneys in adults and children.

The U.S. Department of Housing and Urban Development reports that lead hazards exist in 500,000 homes occupied by young

Universal Wastes

The EPA has established special rules for managing certain wastes that could be hazardous if not properly managed but contain substances that can be recovered. These special wastes, called universal wastes, include certain batteries containing heavy metals (e.g., nickel or cadmium); thermostats that contain mercury; fluorescent lamps that contain mercury; and certain pesticides.

children. The lead found in painted walls of older homes can be very harmful to young children, because the paints may contain very large amounts of lead.

MERCURY

Mercury (Hg) is a poisonous, liquid heavy metal used to make thermometers, barometers, batteries, lamps, skin care products, and medicinal products. Mercury compounds are also used to make neon lights, fungicides, paints, plastics, and electrical equipment such as switches, as well as to produce paper and paper goods.

Exposure to mercury may result from breathing contaminated air, having contact with contaminated soil, or drinking contaminated water. Short-term exposure to high levels of metallic mercury in the air can cause skin rashes and has harmful effects on the lungs and eyes. Long-term exposure to high levels of mercury can cause permanent damage to the brain and kidneys, and it can also be harmful to a developing fetus.

BENZENE

Benzene (C_6H_6) exists naturally in petroleum, gasoline, and tobacco. It is a liquid hydrocarbon that is flammable and toxic to humans and other organisms under some conditions. Long-term exposure may result in stomach irritation, dizziness, rapid heart rate, convulsions, coma, or even death.

Benzene in the air, water, or soil may threaten the health of humans and other organisms. Benzene is released into the air in emissions from coal- and oil-burning power plants, in motor vehicle exhaust, and in tobacco smoke.

The DHHS lists benzene as a carcinogen. Long-term exposure to high levels of airborne benzene can cause leukemia. Automobile mechanics exposed to benzene, a gasoline additive, can suffer from high rates of anemia. Other health problems associated with short-term

TABLE 5-2 Hazardous Wastes in Landfills

Hazardous wastes	All landfilled hazardous wastes (%)
Electroplating wastewater treatment sludge	16.3
Lead	5.9
Chromium	5.9
Electric steel furnace sludge	4.4
Petroleum refinery wastes	3.8

Source: U.S. Environmental Protection Agency.

exposure to high levels of benzene include drowsiness, dizziness, headaches, tremors, confusion, and unconsciousness.

Cadmium

Cadmium (Cd) is a highly toxic heavy metal present in paints, electroplating materials, nickel-cadmium storage batteries, pesticides, and alloys. Cadmium is also used in the control rods of nuclear reactors. Cadmium may enter soil or water during zinc mining or be present in ash formed through the incineration of waste products that are transported to landfills. Chronic effects of cadmium exposure in humans may include kidney or heart damage, decreased fertility, and changes in physical appearance or behavior.

Chromium

Chromium (Cr) is a steel-gray heavy metal that is naturally found in rocks, soil, animals, and volcanic dust and gases. Large amounts of chromium are released into the environment as a result of steel production, chemical manufacturing, and municipal incineration and as a byproduct of sewage sludge. Exposure to chromium happens mostly from breathing contaminated air in the workplace or ingesting water or food from the soil near waste sites. Chromium can cause allergic responses to the skin, and ingesting high doses can cause stomach upset, ulcers, convulsions, lung and kidney damage, or death.

Polychlorinated Biphenyls

Polychlorinated biphenyls (PCBs) are a group of synthetic chemical compounds that can be a health hazard to humans, other organisms, and the environment. In 1979, the sale of PCBs was banned by law in the United States. However, presently PCBs are released into the environment from poorly maintained hazardous waste sites that contain PCBs. The illegal or improper dumping of PCB wastes from electrical transformers containing PCBs into landfills not designed to handle hazardous waste is also another environmental problem. Based on cancer research in animals, the DHHS has determined that PCBs may be carcinogenic.

Perchloroethylene

Perchloroethylene (PCE) or "perc" is a colorless, noninflammable liquid that evaporates easily at room temperatures. It is also known as tetrachloroethylene. PCE is used for dry-cleaning fabrics; for processing and finishing of textiles; as a metal parts degreaser; in the manufacture of other chemicals, primarily fluorocarbons; and as a heat exchange fluid. Perc is also found in small amounts in many household products such as rug and upholstery cleaners. Exposure to this chemical happens mostly from breathing contaminated air and drinking contaminated water.

Short-term exposure to high levels of this chemical can cause damage to the central nervous system and may lead to death.

Love Canal: The First Major U.S. Hazardous Disaster

In 1980, a residential community near Niagara Falls, New York, became the nation's first recognized environmental hazardous waste disaster. The name of the community was called Love Canal. It was built on land that previously held a canal and was used as a dump site by the Hooker Chemical Company. The company, founded in 1905, used the canal to dump about 21,800 tons of chemicals between 1942 and 1952. However, the dumping practice ceased when the city of Niagara Falls began legal proceedings that led to condemnation of the site and a threat to seize the property. To avoid further litigation and to gain release from future liability for problems resulting from the dumping, the company gave its land to the city.

In 1954, the city's school committee bought the site to erect a school building. Houses were also built in the same area. During the winters of 1975 and 1976, disposed drums began surfacing to the top of the site, forming noxious chemicals in outdoor pools and in basements. Over the next few years, residents began complaining to local and state officials about the foul-smelling material oozing from the soil and their basement walls. One of the residents was Lois Gibbs. Lois Gibbs as well as some of her neighbors noticed changes in their family's health. Some of the children were getting sick. Through the

TABLE 5-3 Common Hazardous Wastes

Products	Potentially Hazardous Wastes
Leather	heavy metals, organic solvents
Medicines	organic solvents and residues, heavy metals
Metals	heavy metals, fluorides, cyanides, acid and alkaline cleaners, solvents, pigments, oils
Oil, gasoline, and other petroleum products	oil, phenols and other organic compounds, heavy metals, ammonia salts, acids
Paints	heavy metals, pigments, solvents
Pesticides	organic chlorine and phosphate compounds
Plastics	organic chlorine compounds
Textiles	heavy metals, dyes, organic chlorine compounds, solvents

Source: U.S. Environmental Protection Agency.

residents' activism, tests were conducted on the air, water, and soil qualities at the site. The tests showed high levels of toxic and carcinogenic chemical contaminants, including dioxin. Additional tests conducted by the State Health Department revealed that the Love Canal residents showed higher than expected rates of certain illnesses, including some types of cancer, as well as miscarriages and birth defects. In 1978, Gibbs organized the residents and founded the Love Canal Homeowners Association. The association demanded the evacuation, restitution, and cleanup of the dump site.

Finally in 1978, the New York Health Department ordered temporary evacuation of pregnant women and children under two years of age from the homes immediately next to the canal. Later that year, the governor of New York announced that about 240 families who lived in houses next to canal site were to be evacuated. In 1980, President Carter declared the area a health emergency and announced the evacuation of the rest of the families from the hazardous site.

New York State and the federal government spent more than $40 million to relocate residents, evaluate the damage to the area, and clean up and reclaim the area for use by people. After the cleanup, approximately 200 of the evacuated homes were declared safe for occupancy by the state of New York. Soon after the EPA gave its approval for the homes to be sold to new owners.

Gibbs believes that her major accomplishment was to lead the families of the Love Canal community through two years of struggle to achieve the relocation of all families who wanted to leave. The work of her organization and others resulted in passage of the federal Superfund law.

Chemical Weapons

Special weapons, sometimes referred to as chemical agents, were used by militaries throughout the twentieth century. Some of the chemical weapons consisted of devices that released poisonous or lethal gases intended to debilitate or kill enemy troops. The chemical agents used in such weapons generally caused asphyxiation (oxygen deprivation) or acted on the central nervous system. Modern devices designed to deliver chemical weapons include specially equipped missiles and bombs. Over the years, several governments accumulated stockpiles of unused chemical agents. The aging chemical arsenals presented a potential hazard in the event of a spill or leak of a chemical agent from deteriorating containers. During the last quarter of the twentieth century, many governments took steps to destroy or dispose of chemical weapons. In 1985, the U.S. Congress directed the U.S. Army to destroy its entire stockpile of chemical weapons. The army plans to complete all destruction of chemical weapons by the year 2007. This schedule complies with the international treaty known as the Chemical Weapons Convention (CWC).

Laws to Manage Hazardous Wastes

Superfund

The Superfund is a common name for the Comprehensive Environmental Response, Compensation, and Liability Act (CERCLA) of 1980. The goal of the Superfund Act is to identify and clean up the worst hazardous and toxic waste sites areas on land and water that threatens human health and the environment. The EPA is in charge of the Superfund program.

Years ago, people were less concerned about how the dumping of hazardous wastes might affect their health and the environment. Wastes were dumped and left in open fields, where they contaminated soil and groundwater or were washed away into rivers and streams.

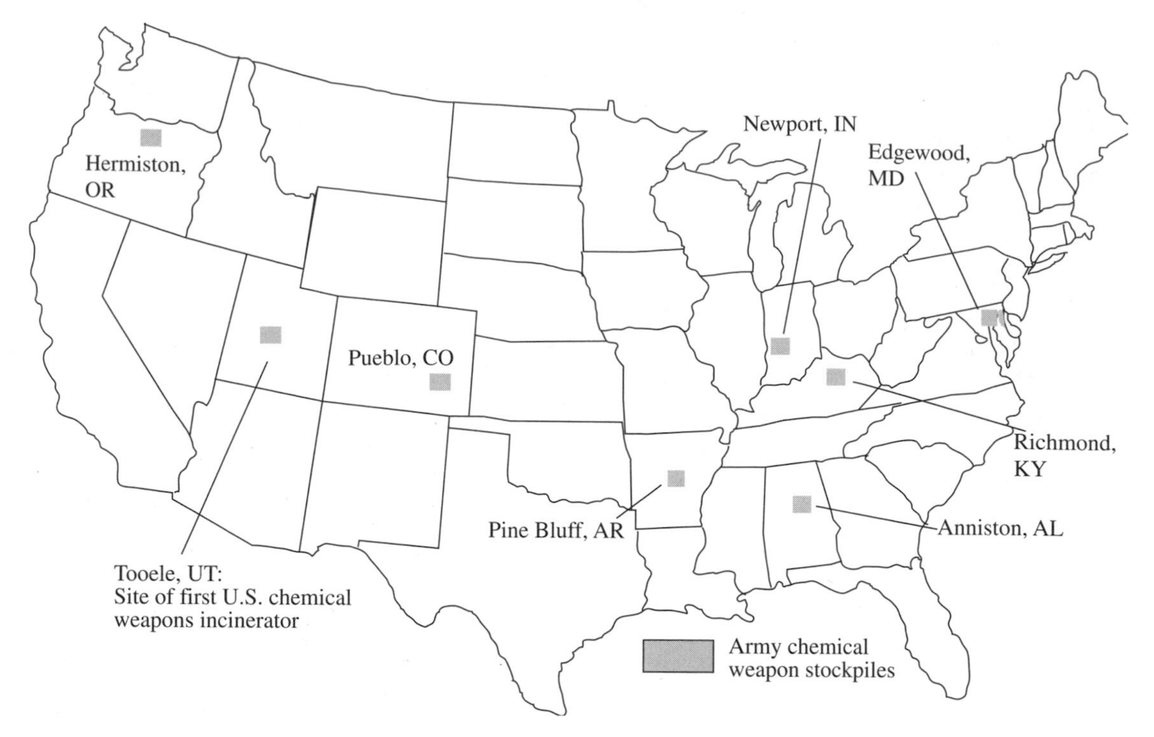

FIGURE 5-1 • United States Army Chemical Weapon Stockpiles There are several chemical weapons storage sites in continental United States and another one on the Johnston Atoll, an island southwest of Hawaii. By 2004, the United States plans to eliminate more than 3 million chemical bombs, rockets, munitions and mines.

Hazardous wastes were discarded in abandoned warehouses and factories, vacant buildings, and open dumps. Hazardous waste was also illegally discarded in remote areas and desolate places. The problems became so dangerous that laws were passed to clean up the hazardous sites.

Under Superfund, the EPA can make responsible individuals and companies or potentially responsible parties pay for cleanup work at Superfund sites. Since 1980, some old hazardous waste sites have been cleaned up and restored to productive use. However, hundreds and hundreds of sites still need to be cleaned up.

Toxic Substance Control Act

Legislation, called the Toxic Substance Control Act (TSCA), enacted by Congress in 1976 required the testing, regulation, and screening of toxic materials such as lead, PCBs, and new chemicals. The testing of the toxic materials determines the potential adverse affects to human health and other organisms, and the environment. Thousands of new chemicals are developed each year, many having unknown toxic or dangerous characteristics. The TSCA requires that any chemical to be sold in the consumer marketplace be tested for possible toxic effects prior to commercial manufacturing. Existing chemicals that are known to pose health or environmental hazards are tracked and reported under the act.

Disposing of Hazardous Wastes

There are several methods for managing and disposing toxic wastes. They include secured landfills, incineration, source reduction, recycling, and deep-well disposal.

Secured Landfills

Hazardous and nonhazardous wastes are deposited into secured landfills monitored by the Resource Conservation and Recovery Act (RCRA). Secured landfills differ from municipal or sanitary landfills, which are intended for household and nonhazardous waste. Unlike most landfills, secured RCRA landfills are designed for the safe disposal of hazardous waste. The secured RCRA landfill system consists of a minimum of two heavy synthetic membranes up to 80 millimeters thick installed over compacted clay at least five feet in depth. Special *leachate* collection systems and ground-monitoring devices protect the surface and groundwater from contamination. All leachate collected from the landfill undergoes treatment.

Incineration

Incineration is a burning process using controlled high temperatures to reduce waste materials to noncombustible ash, carbon dioxide, and water. Incineration is used for two types of wastes—municipal solid waste and hazardous waste. As mentioned in Chapter 4, most municipal solid waste incinerators burn such wastes as garbage, paper, and cardboard after other materials such as rubber, metal, and glass are removed. These incinerators can reach about 500°C (915°F).

To burn toxic wastes, particularly *organic solvents*, PCBs, and chlorinated hydrocarbons, high-temperature incinerators are used for the combustion of these materials. Temperatures during the combustion process can reach 1,000°C (1,830°F) or higher. There are several hazardous waste treatment incinerators in the United States.

The advantages of incineration are that it reduces the volume of original waste by 50 to 75 percent and vaporizes the liquid wastes. The remaining ash is periodically removed, transported, and deposited into sanitary landfills. Emissions from chimney stacks are the major disadvantage of incineration. Toxic ash, sulfur, and nitrogen oxides emitted into the atmosphere contribute to air pollution and acid rain. However, as a consequence of Clean Air Act legislation, many modern incinerators have scrubbers or electrostatic precipitators to remove and trap pollutants before they are emitted into the atmosphere. Another environmental concern is that the ash residue, usually disposed of in landfills, can contain hazardous materials such as heavy metals that can leach into the groundwater.

Environmental groups report that incineration is not necessarily a viable method for disposing of hazardous wastes. Environmental

DID YOU KNOW?

Some countries even export their toxic wastes to other countries. According to Greenpeace, Australia, Canada, Germany, the United States, and the United Kingdom shipped more than 5.4 million metric tons of toxic wastes to countries in Asia between 1990 and 1993. In 1995, however, a global treaty barring rich countries from dumping toxic waste in the Third World was signed.

Refer to Chapter 4 for more information about the Resource Conservation and Recovery Act.

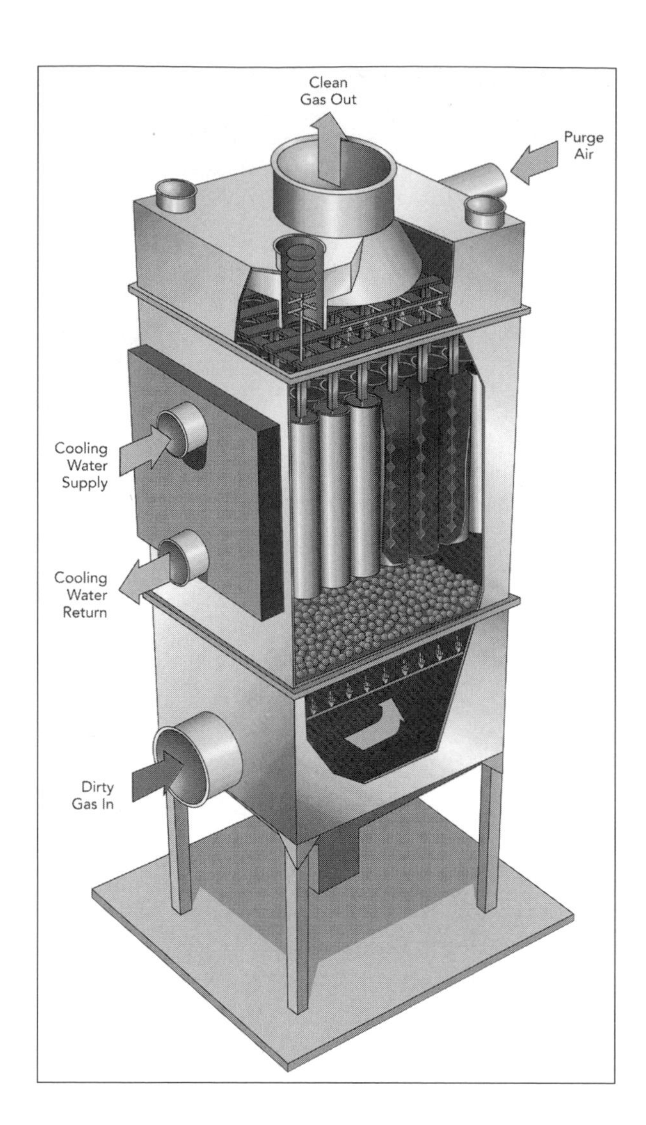

FIGURE 5-2 • The electrostatic precipitator takes in dirty gas and uses electrostatic forces to remove particulates from gas streams. The particulates settle to the bottom where they are removed. (Courtesy of Monsanto)

Before: A former contaminated brownfield is being cleaned up by developers to make way for a marina, shops, and offices. (Courtesy of the Rhode Island Department of Environmental Management)

After: A cleaned up brownfield was redeveloped into a marina with offices and shops. (Courtesy of Courtesy of the Rhode Island Department of Environmental Management)

groups emphasized that source reduction and recycling are better alternatives to waste reduction than incineration. However, incineration for heat recovery such as the burning of waste oil as a fuel is a type of waste reclamation.

Source Reduction and Recycling

As mentioned in Chapter 4, source reduction and recycling prevent or divert materials that would normally be thrown away in a landfill or burned in an incinerator. Source separation takes place at the point of the source where the hazardous waste is produced. The process removes all designated recyclable materials from the waste stream. Source reduction involves separating and isolating hazardous waste from other industrial waste to reduce the volume of hazardous waste.

Brownfields

A brownfield is a term that usually refers to an undesirable industrial property to own or use because contamination exists on the property. Old abandoned industrial properties such as factories may be termed brownfields because the soil or groundwater on the property may contain hazardous materials. Brownfields pose a threat to human health and safety to the environment. Often such properties are not utilized to their best potential. When industrial properties are abandoned, the local government cannot obtain taxes from the unused properties, which creates a burden for the community. Even though there may be interest in purchasing and redeveloping such properties, the prospective purchasers and bankers often fear the consequences of the Superfund Act and other state enviromental acts. These acts could make the prospective owners legally responsible for cleanup costs. In the 1990s, state and local governments implemented new brownfields laws that helped businesses and other prospective redevelopers to investigate, clean up, and purchase brownfields properties by removing some of the Superfund restrictions.

Recycling reduces problems associated with hazardous waste disposal by decreasing the volume of materials that must be deposited in landfills or disposed of in some other way. Nonrenewable resources, land available for landfills, or other means of waste disposal are in limited supply. Thus, finding ways to reduce the overall amount of materials that need to be discarded through source reduction and recycling can help to conserve land space.

DEEP-WELL INJECTION

Deep-well injection is a method of hazardous waste disposal in which wastes are pumped into concrete or steel-encased shafts located deep beneath Earth's surface. Currently deep-well injection is the most widely used method for disposing of hazardous wastes on land.

Despite occasional problems, deep-well injection has been safely used for waste disposal for more than 30 years. The process is usually used for liquid wastes that have been treated to immobilize or destroy their hazardous contents. Before an injection site is selected, geological studies are conducted in the proposed area to ensure that deposited wastes cannot leach from the site and migrate to the surface or contaminate groundwater. An ideal site is one that will trap wastes in the pores of a *permeable* rock layer situated between impermeable rock layers, with all layers located well below the water table. The disposal site must also not be located in a geologically unstable area, such as along a fault or near a region of volcanic activity.

NUCLEAR RADIOACTIVE WASTES

In 2002, some 440 global nuclear reactors will create more than 11,000 tons of radioactive wastes. Radioactive or nuclear wastes are the most hazardous and toxic of all wastes. They include radioactive material from nuclear power plants, nuclear weapon facilities, hospitals, and research centers. Radioactive waste consists of low-level and high-level radioactive materials.

Low-Level Radioactive Wastes

Low-level radioactive waste are tools, clothing, rags, papers, filters, equipment, soil, and construction rubble that were contaminated with low levels of radioactivity. Such material generally originates from hospitals and research centers, as well as from the waste materials left over from mining activities. Low-level wastes are less hazardous than high-level radioactive wastes. Low-level wastes can subside to harmless levels through radioactive decay in only a few years. High-level wastes may take thousands or millions of years to decay to harmless materials.

The *Nuclear Regulatory Commission* (NRC) has developed a classification system for low-level waste. The three general classes of waste are A, B, and C. Class A waste contains lower concentrations of radioactive material than Class C waste. Most low-level wastes are put into drums and buried at commercial disposal sites at the bottom of special trenches. When full, the trenches are covered with clay and top soil.

Low-level waste disposal facilities must be licensed either by the NRC Commission or by state governments in accordance with health and safety requirements. The facilities are to be designed, constructed, and operated to meet safety standards. The operator of the facility must analyze how the disposal plant will perform for thousands of years into the future.

High-Level Radioactive Wastes

High-level radioactive wastes are extremely toxic and more hazardous than low-level radioactive wastes. Unlike low-level wastes, which have a short half-lives, high-level radioactive wastes are usually dangerous for thousands or even millions of years. Dangerously radioactive wastes come from two sources. One source includes spent *fuel rods* from nuclear reactors that are no longer able to produce nuclear fission. The second source includes liquid and solid wastes from plutonium production. Plutonium is produced inside a nuclear reactor and is the primarily fuel in making nuclear weapons.

When spent fuel rods are removed from a nuclear reactor, they are put into a pool of water and stored temporarily at the nuclear reactor site. The water in the pool acts as a radiation shield and coolant. Some of the spent fuel is also placed in dry canisters or casks made of metal or concrete. However, storing the spent fuel in pools is intended only as a temporary measure until a permanent disposal site is found.

Sources of Radioactive Wastes

Transuranic Wastes

Transuranic Wastes are byproducts from defense programs produced during nuclear weapons research and fabrication and reactor fuel assembly. Transuranic wastes include contaminated equipment, tools, protective

A technician is at the vault door of a waste storage area used for low-level radioactive solid wastes. The storage areas are divided into a transuranic storage area and a subsurface disposal area. (Courtesy of the U.S. Department of Energy.)

TABLE 5-4 Spent Fuel Stored at U.S. Nuclear Plants (metric tons)

State	1995
Alabama	1,439
Arizona	465
Arkansas	581
California	1,391
Colorado	15
Connecticut	1,254
Florida	1,440
Georgia	1,019
Illinois	4,292
Iowa	235

Several nuclear power plants in the United States are storing spent fuel. There are approximately 52,000 tons of spent fuel from commercial, research, and military reactors and millions of gallons of radioactive waste from plutonium processing.
Source: National Renewable Energy Laboratory.

clothing, glassware, soils and sludge, and other materials used in laboratories and research centers. Transuranic wastes are contaminated with industrial radioisotopes heavier than uranium. Because these wastes decay slowly, they require long-term waste storage. Most of the transuranic waste can be packaged and stored in metal drums or in metal boxes.

URANIUM MILL TAILINGS

Other radioactive wastes include uranium mill tailings from the mining and milling of uranium ore. The tailings consist of rock and soil containing small amounts of radium and other radioactive materials. Uranium mill tailings become a radioactive waste disposal problem because radon, a radioactive gas, is produced when radium decays.

Radioactive Waste Disposal

Most experts believe that the need to find safe and permanent disposal of nuclear waste is becoming more critical because storage pools are almost full at some nuclear power plants. One area that may be used as a place to store nuclear waste is in the Yucca Mountain in Nevada.

YUCCA MOUNTAIN PROJECT

The U.S. government is responsible for finding a place to safely dispose of this spent nuclear fuel. In 1982, Congress passed the Nuclear Waste Policy Act that directed the U.S. Department of Energy to find a site. Although sites in Texas and the state of Washington were considered,

The proposed Yucca Mountain repository is a high-level nuclear waste site that could store approximately 75,000 tons of wastes. The photo shows the south portal of the Exploratory Studies Facility. (Courtesy of the U.S. Department of Energy)

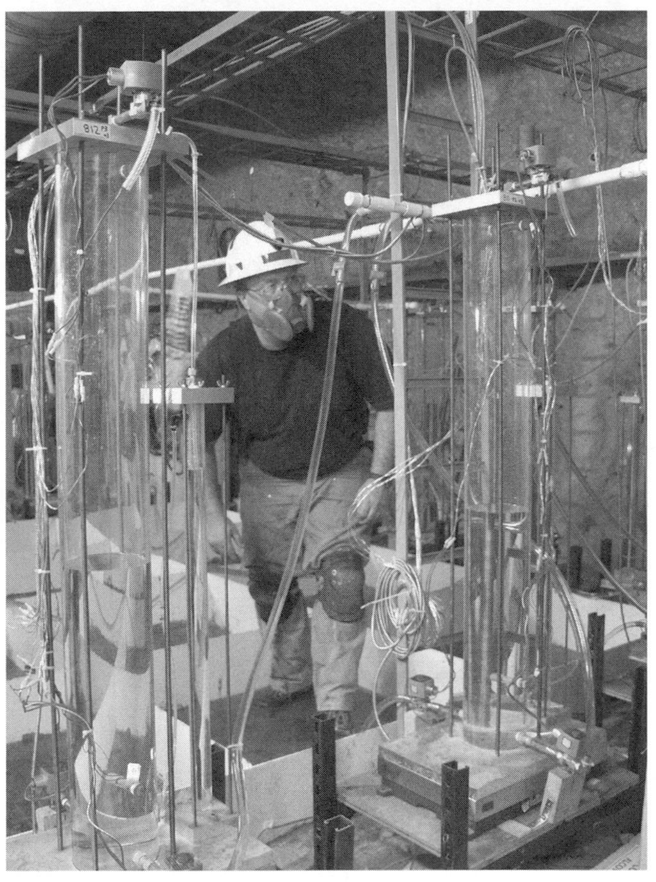

Deep inside the proposed Yucca Mountain repository, electric locomotives will carry radioactive disposal containers through some 80 kilometers of tunnels. The photo shows an inside view of the Yucca Mountains. A project scientist is testing for water movement in the rock. (Courtesy of the U.S. Department of Energy)

Congress selected the Yucca Mountain in Southwestern Nevada. The program is called the Yucca Mountain Project, which will be a long-term nuclear facility. It can be used as a *repository* for storing the nation's commercial and defense spent fuel and high-level radioactive wastes (HLW). The project remains under the supervision of the Department of Energy.

According to project researchers, the Yucca Mountain was a good selection as a long-term containment facility because it is located in a remote area and a good distance, about 160 kilometers (100 miles), from Las Vegas. Nevada also has a very dry climate, less than 15 centimeters (six inches) of rainfall per year. The location has a deep water table of 260 to 330 meters (800 to 1,000 feet) below the level of the potential repository site.

However, the Yucca Mountain Project is controversial. Several environmentalists and organizations have voiced their objections to the project because they believe the stored radioactive wastes could be a health risk and a danger to future populations in the area. They point out that earthquakes could damage and break up the stored containers, in which case radioactive wastes could leak into the groundwater. They are also worried that accidents could occur when tons of radioactive wastes are being transported by trains and trucks through populated areas on their way to the Nevada site.

HANFORD NUCLEAR RESERVATION

In 2005, the Hanford Nuclear Reservation will begin operation as a radioactive waste treatment plant to clean up radioactive wastes. The Hanford Nuclear Waste site is an area of approximately 1,450 square

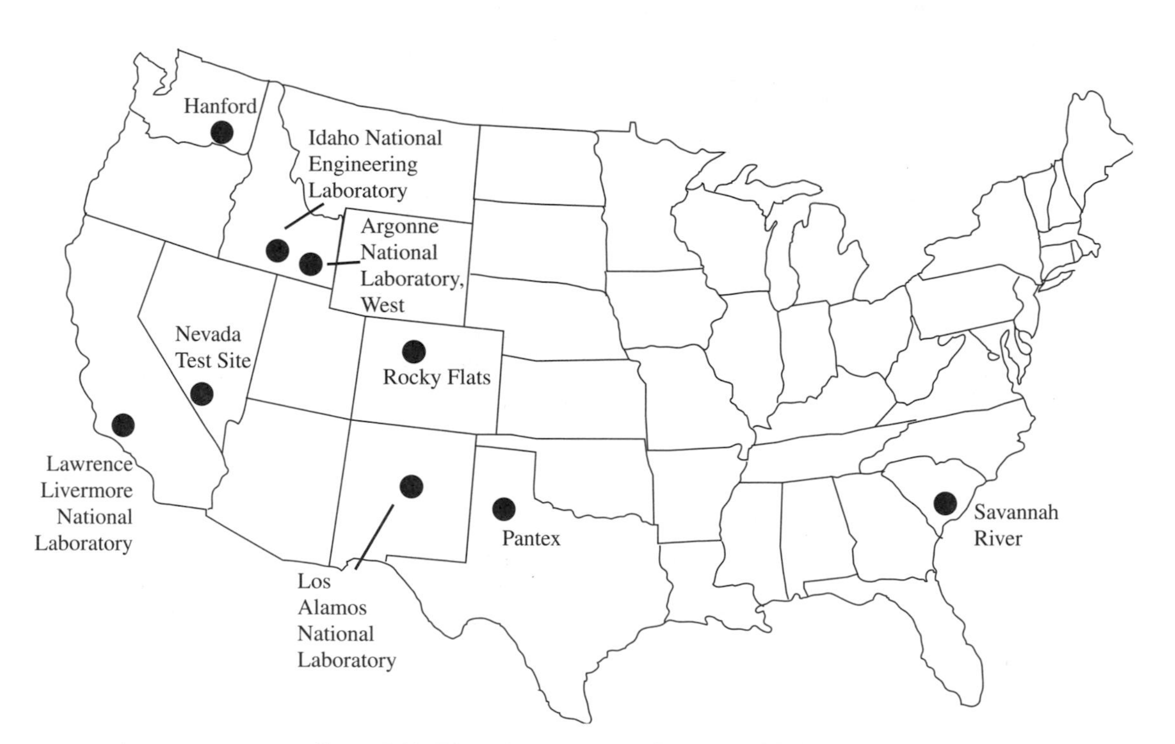

FIGURE 5-3 • Plutonium Storage Sites One of the largest laboratories ever built for plutonium (Pu-239) production was the Hanford site in the state of Washington. Hanford reactors made plutonium for the first nuclear explosion in New Mexico in 1945. The Hanford site contains the largest source of high-level nuclear wastes. Presently, more than 11,000 workers are involved in cleanup tasks at the Hanford complex.

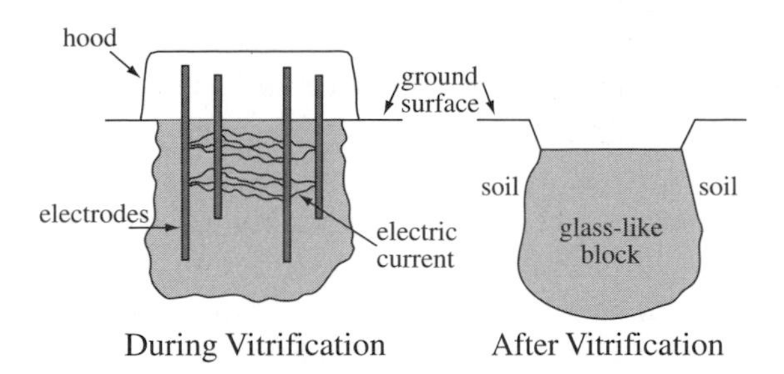

FIGURE 5-4 • Vitrification is a technology that can be used for hazardous and radioactive solid waste as well as for other wastes. Approximately 53 million gallons of radioactive waste in Hanford are scheduled for vitrification.

kilometers (560 square miles) located in the southeastern part of Washington State near the Columbia River. The Hanover Nuclear Waste site was used at one time by the U.S. Department of Defense for the production of plutonium fuel, a radioactive material used for developing nuclear weapons.

The production of nuclear weapons began in the 1940s during World War II, when little consideration was given to the management of nuclear wastes. Between 1944 and 1989, the Hanford Nuclear Waste Site produced over 50 tons of weapons-grade plutonium, which represented more than half of the U.S. supply. During that time, much radioactive waste was released into the environment in and around the Hanford facility.

DID YOU KNOW?

Plutonium developed at the Hanover Nuclear plant during the 1940s was used in the atomic bomb that was dropped on Nagasaki, Japan, in World War II.

The new Hanover radioactive treatment plant will use a process called vitrification. Vitrification is a process in which radioactive waste materials are mixed with molten glass to form solid glass blocks. To make the glass blocks, the radioactive waste and the glass-forming materials are mixed together and then melted at 1,000°C (2,000°F). The molten glass is poured into blocks for long-term storage. The radioactive glass blocks containing high levels of radioactive materials will be stored in a national repository. The blocks containing lower levels of radiation will be buried in trenches on the site.

Vitrification does not reduce radioactivity, but it changes the form of the wastes from a leachable liquid to an immobile solid. The final product captures radioisotopes, thus preventing these potentially harmful wastes from contaminating soil, groundwater, or surface water.

Although vitrification provides a means for containing radiation, simply vitrifying a material does not necessarily produce an environmentally stable product. To ensure that a waste material is stable for disposal, the chemistry of the material needs to be studied carefully. Not all radioactive wastes currently produced can undergo vitrification, and additional testing of this process is continuing. The Hanford cleanup plan is to treat all the waste by 2028.

DID YOU KNOW?

Plants such as Indian mustard are used to absorb radioactive materials in soil. Once the plants absorb all the radioactive materials, they are removed and properly disposed of. This process is a type of "phytoremediation," which refers to the use of plants in remedies for wastes in the environment.

Radiation Sickness

Health Effects of Radiation

Overexposure to radiation emissions from such sources as X-rays, gamma rays, and nuclear weapons and radioactive fallout is dangerous. These emissions can damage or destroy cell division, thereby preventing the normal replacement or repair of blood cells, skin, and other tissues. Low levels of radioactive exposure can cause loss of hair, vomiting, and stomach illnesses. Higher levels of radioactive exposures can cause respiratory problems as well as leukemia and bone cancer. Direct exposure to radiation emissions can lead to death within a few days.

Radiation doses are measured in sieverts (Sv) with 0.05 Sv being the maximum radiation dose the human body should absorb in one year. Radiation sickness can occur when the body absorbs between 1 to 2 Sv. Radiation sickness has occurred from mining of uranium, from nuclear weapon production, and from accidents at nuclear power plants.

Nuclear Power Plant Accidents

Probably the most well known nuclear power plant accident occurred at the Chernobyl power plant in Ukraine. In 1986, fire and explosion at the power plant created nuclear fallout and radiation that contaminated large areas of northern Europe, particularly the United Kingdom, Finland, and Sweden. Some reports indicated that as many as 20 countries received high levels of fallout from the accident. The initial fire, which lasted for several days, killed about 30 people; but a larger number of others are expected to die from prolonged exposure to the resultant radiation. Health experts expect to see a rise of cancer deaths over the next 50 years. They project that the total may be as high as 40,000 cases. Most of these deaths will occur in the immediate vicinity

TABLE 5-5 World Usage of Nuclear Energy

Rank	Country	Electricity from nuclear generators (% of total)
1	Lithuania	77
2	France	76
3	Belgium	55
4	Sweden	46
5	Ukraine	45
6	Slovakia	44
7	Slovenia	38
8	Bulgaria	42
9	Republic of Korea	41
10	Switzerland	41

All nuclear-energy producing countries, such as those listed above, need plans to store and dispose of radioactive wastes from their nuclear plants.

of the nuclear reactor. However, no one really knows how many will perish as a result of radiation sickness. Another health concern is the increased number of people who have been diagnosed with tuberculosis since the accident.

The Chernobyl explosion occurred during a test when fuel rods became so hot that they caused a steam explosion. The explosion blew off the top of the reactor, ejecting radioactive fuel and burning control rods into the atmosphere. Risking their own lives, lab technicians, firefighters, and others worked courageously to control the fire. Several sacrificed their lives, and many were commended for their bravery. Today the nuclear power reactor has been closed and sealed with concrete. It is visited by researchers to learn more about the effects of nuclear fallout and its impact on people and the environment.

FUTURE PROBLEMS

Safely disposing of radioactive wastes that may remain radioactive for thousands of years is a top priority for many countries that use nuclear energy. According to the International Atomic Energy Agency, about 10,000 cubic meters (350,000 cubic feet) of high-level waste accumulates each year. Unfortunately, many countries using nuclear energy have no long-term plan or program for storing radioactive wastes safely.

Vocabulary

Carcinogens Substances that produce cancer.

Fuel rods Nuclear fuel in the form of a rod used in a nuclear reactor.

Leachate Fluid containing substances that are washed out of the soil.

Nuclear Regulatory Commission A U.S. government organization that issues licenses for the construction and operation of nuclear power plants.

Organic solvent A liquid, such as benzene, that is used to dissolve materials such as grease, varnishes and paints.

Permeable Allows water to pass through material.

Repository A place where nuclear waste can be stored.

Activities for Students

1. Do you live close to an industrial area? Investigate how these industries are disposing of their wastes. Are they using environmentally friendly methods?
2. Create a presentation for people in your community to raise awareness of risks associated with toxic wastes and to show how they can protect themselves.
3. Investigate how different countries are disposing of their nuclear waste. Should there be international regulations in place to protect the environment, or should individual countries be responsible for setting standards? How should the standards be enforced?

Books and Other Reading Materials

Altman, Roberta. *The Complete Book of Home Environmental Hazards.* New York: Facts on File, 1990.

Chernousenko, V.M.M. *Chernobyl: Insight from the Inside.* New York: Springer-Verlag, 1992.

Cohen, Gary, and John O'Connor. *Fighting Toxics: A Manual for Protecting Your Family, Community, and Workplace.* Washington, D.C.: Island Press. 1990.

Haun, J. William, and Bill Haun. *Guide to the Management of Hazardous Waste: A Handbook for the Businessman and Concerned Citizen.* Golden Colo: North American Press, 1991.

Rogovin, Mitchell, and George T. Frampton. *Three Mile Island: A Report to the Commissioners and to the Public.* Vols. I–II. Washington, D.C.: Nuclear Regulatory Commission, 1980.

Wagner, Travis P *The Complete Guide to Hazardous Waste Regulations: RCRA, TSCA, HMTA, OSHA, and Superfund.* New York: John Wiley & Sons, 1999.

Websites

Brownfields Projects, http://www.epa.gov/brownfields/

Chemical Stockpile Disposal Project (CSDP), http:// www-pmcd.apgea.army.mil/graphical/CSDP/index.html

Environmental Defense Fund, The organization provides access to data on wastes and chemicals at United States sources, http://www.scorecard.org

International Atomic Energy Agency, "Managing Radioactive Waste" Fact Sheet, http://www.iaea.org/worldatom/inforesource/factsheets/manradwa.html

International Atomic Energy Agency International Chernobyl Assessment Project, http://www.iaea.org/worldatom/inforesource/other/chernobook/index.html

Landfills: An Issue Confronting Our Sustainable Use of the Land, http://www.lalc.k12.ca.us/uclasp/issues/ landfills/landfills.htm

Landfills—Solid and Hazardous Waste and Groundwater Quality Protection, http://www.gfredlee.com/landfill.htm

Material Safety Data Sheets (MSDSs), which contain health and safety information about chemicals and products, are available online at the *Vermont SIRI MSDS Archive*, http://www.siri.uvm.edu/msds/grep/g2.cgi

National Research Council, Board on Radioactive Waste Management, http://www4.nas.edu/brwm/brwm-res.nsf

OECD Nuclear Energy Agency, Chernobyl Executive Summary, http://www.oecdnea.org/html/rp/chernobyl/c0e.html

Superfund Sites: Find Your State, http://www.epa.gov/superfund/sites/npl/npl.htm

Teach with Databases, Toxic Release Inventory, http://www.nsta.org/pubs/special/pb143x01.htm

U.S. Department of Energy, Office of Civilian Radioactive Waste Management, http://www.rw.doe.gov

U.S. Department of Health and Human Services, Agency for Toxic Substances and Disease Registry (ASTDR), www.atsdr.cdc.gov

U.S. Environmental Protection Agency Superfund Program, http://www.epa.gov/oerrpage/superfund/programs/er/hazsubs/index.htm

U.S. Environmental Protection Agency Superfund Program Home Page, http://www.epa.gov/superfund/index.htm

U.S. Nuclear Regulatory Commission, Radioactive Waste Page, http://www.nrc.gov/NRC/radwaste

U.S. Occupational Safety and Health Administration (OSHA), http://www.osha-slc.gov/sltc/hazardoustoxicsubstances/index.html

Yucca Mountain Project, http://www.ymp.gov/

CHAPTER 6

Loss of Forests and Agricultural Pollution

The world's forests covers almost one-third of Earth's land surface, and they are important natural resources for people. Each year, millions of hectares of forest are cut down for timber production, fuelwood, and land used for housing, farming, ranching, and mining activities. Forestry, mining, farming, and ranching are an important part of the economy. They provide us with food, shelter, employment, and energy. However, excessive logging, land degradation by mining, overgrazing of livestock, and poor agricultural managing practices are causing environmental problems. Some of these problems include water and wind erosion, air and water pollution, the depletion of nutrients in the soil, the wearing away of land to desert conditions, and the loss of habitat.

DEFORESTATION

The temporary or permanent removal of trees and vegetation for commercial purposes is called deforestation. About 50 percent of the world's original forest cover is already lost to logging activities, forest fires, land clearing for farms and ranches, and other human activities.

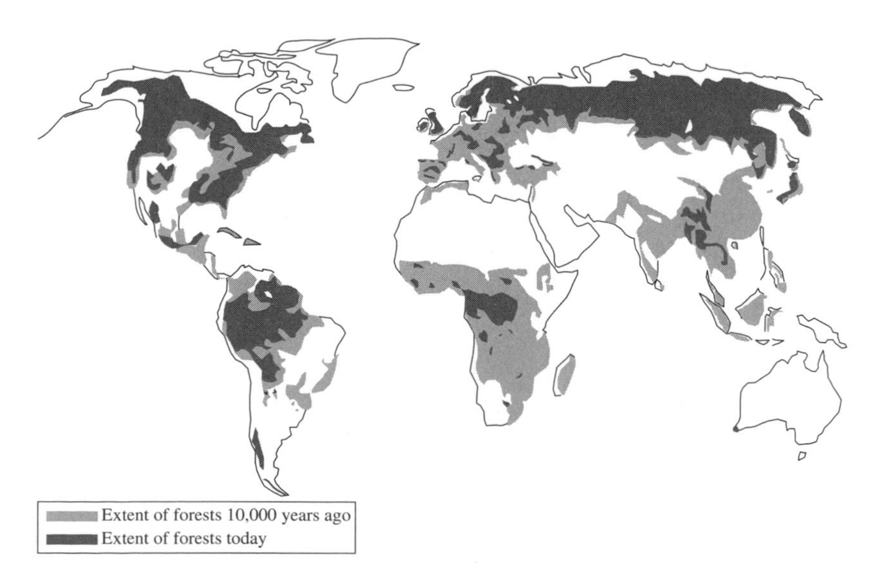

FIGURE 6-1 • Disappearing Forests

As the human population grows, more land, trees, minerals, and other natural resources are needed. In many places in the world, forests continue to disappear at an alarming rate. Over 8 billion hectares (about 19 billion acres) of forest existed in the world 8,000 years ago; today only 3–4 billion hectares (about 10 billion acres) remain. The pace of forest destruction accelerated in the 1990s and continues to rise. According to environmental sources, over 400,000 hectares (950,000 acres) of forest are cleared or degraded every week.

Deforestation is most severe in the tropical rainforests of Africa, Asia, Central America, and South America. About 18 million hectares (45 million acres) of tropical rain forests are cleared annually. Over half of the world's original tropical rain forests have already been destroyed by deforestation. The remaining rainforests, almost 1.8 billion hectares (4.5 billion acres), cover only 5 to 7 percent of the world's land. As an example, Brazil loses 15 million square kilometers (5.7 million acres) of forests a year, more than any other country. At the current rate of deforestation, scientists predict that no rainforests will remain by the middle of the twenty-first century.

Refer to Volumes I and II for more information about forests and rainforests.

Causes of Deforestation

Different forestry practices are used to cut down and harvest trees. Two of the practices that can lead to deforestation are slash-and-burn and clear-cutting activities.

Slash and Burn

A major practice that involves the cutting down of large numbers of trees from a forest to clear the land for agricultural uses is called slash and burn. This process liberates nutrients and organic matter otherwise bound up in the vegetation, which enriches the soil for cultivation or livestock grazing. Slash-and-burn practices are most common in regions where the dominant *biome* is a tropical rainforest, such as in Southeast Asia, South America, and some regions of Africa. In regions where slash-and-burn methods are used, the practice often is accompanied by

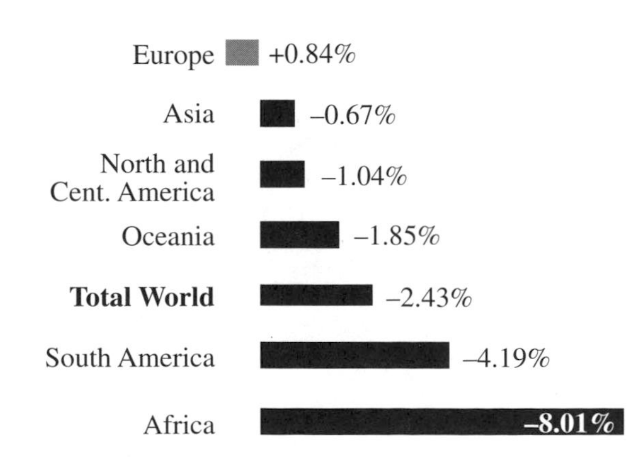

Figure 6-2 • Forest Loss by Region: Percentage Change, 1990–2000 Except for Europe, most continents have had forest losses since 1990.

shifting cultivation. A shifting cultivation is an agricultural process in which plots of land that have been cultivated for only a few years are abandoned while new plots are created through the additional cutting down and burning of trees. Through the process of *ecological succession*, the abandoned area is eventually taken over by weeds and later bushes and trees. Erosion can also be severe near waterways and slopes.

Slash-and-burn activities are also used to clear the land for cattle ranching. In fact, cattle ranching and other types of livestock production account for a significant amount of deforestation in South and Central America. In these areas, some 52,000 square kilometers

Slash and burn is a forestry practice that involves cutting down large tracts of trees to clear the land for farming and ranching activities. (Courtesy of Rainforest Action Network)

(20,000 square miles) of forest—an area the size of New Jersey—are cleared each year to create grazing pastures for cattle.

CLEAR-CUTTING

Refer to Volume II for more information about how paper is made.

Clear-cutting activities are used primarily for cutting down trees for timber, paper pulp, and other wood products or to establish tree farms and plantations. Trees also removed to provide fuelwood for many people. About 1.5 billion people in developing countries depend on fuelwood as their major source of fuel.

Clear-cutting is a tree-harvesting method of cutting down all trees and vegetation in an assigned area. Like slash-and-burn activities, clear-cutting is a controversial subject for many environmentalists. Some foresters like the process because it is easy to cut an entire forest and then replant trees. They believe that the cleared land is ideal for the regeneration of the shade-intolerant tree species because direct sunlight is needed for seed germination. They further state that in small areas of 10 hectares (25 acres) or less where the ground is level and precipitation is light to moderate, clear-cutting is a feasible method for harvesting shade-intolerant trees.

However, clear-cutting over vast areas can cause destruction of habitats, heavy soil erosion, and deterioration of soil. In tropical forests, clear-cutting is especially a problem in areas where tree species are poorly suited to *regeneration*. When the poor soil of the tropical forest is exposed to hot temperatures, the heat accelerates the decay of organic matter in the soil and reduces the soil's moisture and nutrients that are important for tree growth. Important nutrients in the soil are also lost

A clearcutting operation in Oregon; clearcutting is a harvesting method of cutting down all trees and vegetation in an area. (Courtesy of United States Department of Agriculture)

to runoff during rains. Landslides can occur when clear-cutting is practiced along unstable, hilly slopes in locations with heavy rainfall. Landslides develop because there are no trees and other vegetation to hold the soil in place. Once the vegetation is removed, most of the roots of the cut trees do not regenerate but decay, losing the ability to hold the soil in place. Sometimes forests that are clear-cut do not regenerate. Clear-cutting may also prevent the original ecosystem from being restored in that area. Today clear-cutting is being evaluated more closely as it relates to the surface and soil of the land, the number of trees to be cut, and the kinds of trees on the land.

ALTERNATIVES TO SLASH AND BURN AND CLEAR-CUTTING

Alternatives to slash and burn and clear-cutting include selection cutting, and strip cutting. In selection cutting, only the mature trees are marked and cut down; the immature ones are left to grow. In this method, the remaining trees of different sizes and types are left to maintain the ecosystem. In strip cutting, trees are cut in narrow rows or strips. The uncut areas remain open for use in recreation and for wildlife habitats. The trees that are not stripped provide seeds and protect young trees from the sun and wind.

Refer to Volume V for more information about sustainable forestry practices.

Consequences of Deforestation

CHANGES IN BIOLOGICAL DIVERSITY

Deforestation destroys the habitats of countless animal and plant species. Many species have now become extinct or are being pushed to the brink of extinction because their habitats are being destroyed. More than 8,750 of the 80,000 to 100,000 tree species known to science have been threatened to extinction, and 77 are already extinct. This number includes almost 1,000 species believed to be critically endangered. Fewer than one-quarter of the threatened tree species are protected by conservation measures. Only 12 percent of these species are recorded in protected areas, and only 8 percent are known to be in cultivation. Threats to tree species include cutback for timber and fuelwood, agriculture, expansion of human settlements, uncontrolled forest fires, invasive exotic or *alien species*, and unsustainable forest management.

CHANGES IN SOIL QUALITY

In many countries, large-scale farms, ranches, and plantations are rapidly replacing tropical rainforests. The deforestation of the rainforests has already caused significant changes in soil quality. Despite the abundance of vegetation, tropical rainforests have surprisingly poor soil. When trees are cleared, the already nutrient-poor soils of the rainforest become even more infertile. The hot tropical sun also bakes exposed soils into a hard, brittle surface that can more easily be eroded by rainfall. After a few years, the soils become destroyed and eroded by the movements of vehicles and cattle. Eventually the land is abandoned, and a new patch of forest is cleared.

CHANGES IN GLOBAL ENVIRONMENTS

Deforestation can also lead to changes in local and global climate. Tropical rainforests help control local climates by maintaining humidity and offering protection from the wind. Water evaporates from the leaves of trees and other vegetation, enters the atmosphere, and eventually falls back to land during heavy rainfalls. When these forests are cleared, this important source of water vapor is also lost. Eventually the loss of water vapor in the atmosphere can lead to drought conditions. The continued deforestation of the world's tropical rainforests may also alter global climate by contributing to the greenhouse effect.

The greenhouse effect, mentioned in Chapter 2, is a natural phenomenon that maintains global temperatures. The greenhouse effect works because several gases, primarily carbon dioxide (CO_2), forms a blanket in the atmosphere and traps heat, much as the glass walls of a greenhouse trap the sun's heat. The loss of trees contributes to the greenhouse effect in two ways. First, when trees are burned, large amounts of carbon dioxide are released into the air. The loss of trees and other plants means that less carbon dioxide is being absorbed from the atmosphere for photosynthesis. Overall, deforestation contributes to about 25 percent of all global carbon dioxide emissions. Scientists report that the greenhouse gases will continue to trap more heat in the atmosphere, thus causing Earth to grow warmer.

Solving the Problems of Deforestation

Currently governments, environmental organizations, and scientists are working on ways to slow the rates of deforestation. Their efforts include the following activities:

- Establishing tropical forest reserves to protect forested land
- Improving forest management of unprotected forests
- Encouraging new agricultural technologies to prevent soil degradation
- Curbing the demand for tropical forest products
- Involving local people in managing the forests, particularly rainforests

One of the most important ways to reduce the rate of deforestation is to implement sustainable forest management practices for harvesting trees. For the survival of forests, it is therefore important for countries to consider exporting and importing only sustainable forest products. One organization that promotes sustainable forest management is the World Wildlife Fund.

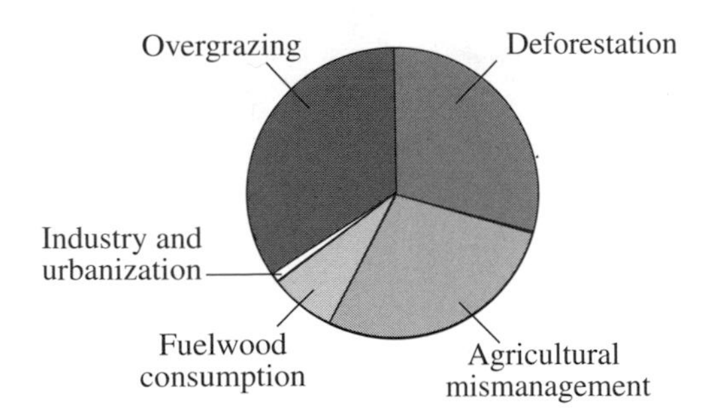

FIGURE 6-3 • Causes of Land Degradation

AGRICULTURAL POLLUTION

Agricultural activities can adversely affect the environment through agricultural pollution. Agriculture pollution is the release of pollutants from farming and ranching activities into the environment. Agricultural pollution may result from two major sources. One source includes the runoff of fertilizers, manure from feedlots, herbicides, and pesticides that pollute waterways. Agricultural runoff is the primary source of pollution in our nation's rivers, lakes, and streams. The other source of pollution is the degradation of topsoil by erosion that results from unsustainable agricultural practices.

Fertilizer Runoff

Chemical fertilizers are added to soil to replace nutrients removed by plants. Fertilizers promote plant growth; however, when washed away into aquatic ecosystems, the excessive fertilizers in the water can cause eutrophication. Eutrophication results from the accumulation of nutrients, especially phosphorus and nitrogen compounds, in a body of water. When too many nutrients are present in an aquatic ecosystem, a population explosion of algae, called an algal bloom, may occur.

As you learned in Chapter 3, during an algal bloom, the algae population can exceed the *carrying capacity* of the ecosystem. When this happens, large numbers of algae suddenly die, stimulating the decomposition process. As large numbers of bacteria and fungi carry out decomposition, they may use so much of the water's dissolved oxygen that the ecosystem becomes unable to support other aquatic organisms.

Feedlot Runoff

Approximately 66 percent of all livestock that is processed for meat products in the United States are raised in feedlots. Some of the largest feedlots can hold up to 50,000 animals (i.e., swine, cattle, chicken, etc.). Animal waste, or manure, is a byproduct of the feedlot. About 23 kilograms (50 pounds) of manure is produced per day per animal. These nitrogen-rich wastes can be washed off the feedlots and into aquatic ecosystems, thereby causing the same problems as the runoff of chemical fertilizers. The dried wastes can also be swept away by wind and

Desertification

Extensive farming, overharvesting of fuelwood, and cattle raising can cause desertification. This occurs when the land is stripped of its plant cover, exposing the topsoil to erosion. The erosion occurs when loose and exposed topsoil is easily carried away by wind, running water, and other agents of erosion. North and South America; Mediterranean Europe; southern and western Asia, including China and Mongolia; Australia; and many countries in Africa suffer from desertification.

Desertification represents problems that are not easy to fix. The land becomes sandy, coarse, and stony, and it loses its capacity to hold water. Other problems of desertification includes the lowering of the water table making it difficult to pump water for crops.

As mentioned earlier, slash-and-burn practices can lead to deforestation, but they can also lead to desertification. In this type of farming, areas of forest may be cleared and then burned to permit planting of crops. The burning produces a highly fertile but short-lived soil that fails to support agriculture after a few harvests. When large areas are cleared in this fashion, rapid soil erosion and the failure of the original forest to regrow trees on the land may result in eventual desertification and the drying up lakes and ponds used for irrigation. There is also an increase of *salinity* in the soil as a result of irrigation history. Native plants are removed that would normally control wind and soil erosion.

In arid or semi-arid regions, desertification may occur as a result of human activities or climatic change. The great deserts of the world have developed in response to long-term climatic trends. However, the extent of deserts has increased in many areas at a more rapid pace than would occur without human influence.

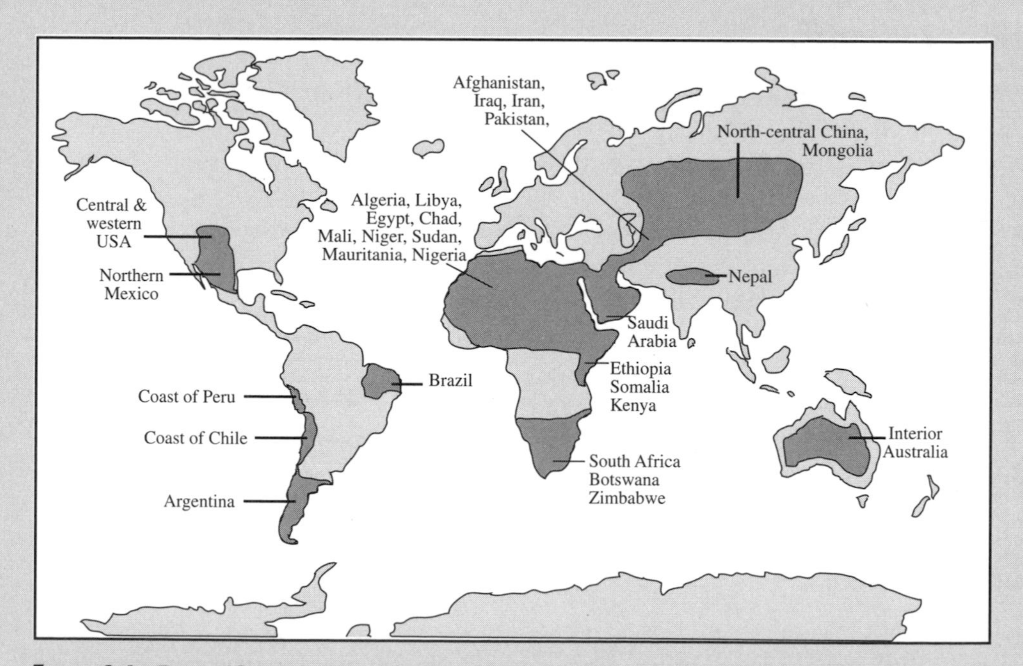

FIGURE 6-4 • Desertification Desertification shown in the dark tint is a major problem in parts of North and South America, Mediterranean Europe, Southern and western Asia including China and Mongolia, Australia, and many countries in Africa.

(*continued*)

Desertification (*continued*)

Commonly desertification occurs on the edges of existing deserts, in dry areas, or at the edges of grasslands and proceeds outward, creating a larger area of desert conditions. This progressive growth of a desert takes place not as a uniform expansion but in increments, one patch followed by another. Droughts can hasten desertification.

Irrigation practices can also lead to desertification. Irrigation can cause the accumulation of salts in poorly drained soil. Over time as this process makes the irrigated soil increasingly unsuitable for farming, erosion of the topsoil can occur. This sequence of changes may ultimately result in desertification.

To help reduce the problem of desertification, farmers often use farming methods such as contour plowing, terracing, strip cropping, and no-till agriculture to reduce the effects of erosion. Ranchers also help reduce erosion by periodically moving grazing animals to different pastures to prevent overgrazing.

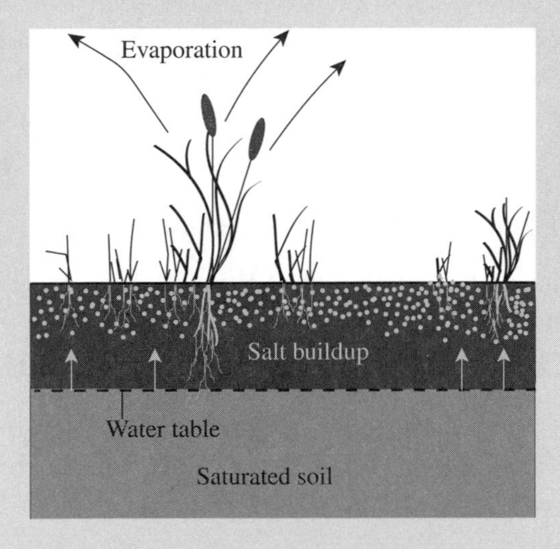

FIGURE 6-5 • Salinization of soil occurs when soluble salts such as sodium chloride, calcium chloride, and magnesium chloride accumulate in soils to a level that harms plants and prevents their growth. Salinization caused the Mesopotamia farmers to abandon their fields after many years of good farming. Salinization is still an environmental problem today in many countries including the United States.

Contour farming is a soil conservation practice in which soil is plowed according to the natural shape of the land to prevent topsoil erosion due to running water and wind. (Courtesy of United States Department of Agriculture)

deposited in waterways. In some cases, manure dust is so strong that it causes the air quality to exceed healthy air pollution standards. Some ranchers and environmentalists are recommending more limits on the location and operation of feedlots.

Pesticides

Pesticides are chemicals used for pest control in agricultural and gardening activities. Pesticides can be highly toxic to humans and other organisms if not used or handled properly. Pesticides are divided into three main groups: insecticides, herbicides, and fungicides. Insecticides are used to control and kill insects. Herbicides are used to control unwanted plants such as weeds. Fungicides are applied to kill fungi and other parasites.

The widespread use and disposal of pesticides by farmers, institutions, and the general public provides many possible sources of pesticides in the environment. After release into the environment, pesticides may have many different final outcomes depending on how they are released. Pesticides that are sprayed can move through the air and may eventually end up in other parts of the environment, such as in soil or water. Pesticides that are applied directly to the soil may be washed off the soil into nearby bodies of surface water or may percolate through the soil to lower soil layers and groundwater; they may even migrate as dust particles in the air.

Break Down of Pesticides in the Environment

Two things may happen to pesticides once they are released into the environment. They may be broken down, or degraded, by the action of sunlight, water, chemicals, or microorganisms such as bacteria. This degradation process usually leads to the formation of less harmful breakdown products. The second possibility is that the pesticide will resist degradation. The resistant pesticides can thus remain unchanged in the environment for a long period of time.

The farmer is applying a chemical control application to crops. Chemicals used for pest control in agriculture and gardening are called pesticides. The three main groups of pesticides include herbicides, fungicides, and insecticides. (Courtesy of Jeff Varuga, USDA, NRCS)

The pesticides that are most rapidly broken down have the shortest time to move through the environment or to produce adverse effects on human health and the environment. In contrast the ones that are broken down slowly have more time to remain in the environment and cause problems. The pesticides that last the longest are called persistent pesticides. They can travel over long distances and can build up in the environment, leading to a greater potential for adverse effects.

In addition to showing resistance to degradation, pesticides have a number of other properties. One property or characteristic of pesticides is that they are *volatile*. The ones that are most volatile have the greatest potential to evaporate into the atmosphere, and if persistent they travel long distances. Another important property is their solubility, or how easily they dissolve in water. If a pesticide is very soluble in water, it is more easily carried off with rainwater as runoff or through the soil as a potential groundwater contaminant. In addition, the water-soluble pesticide is more likely to stay mixed in the surface water, where the chemical agent can have adverse effects on fish and other organisms. If the pesticide is very insoluble in water, the chemical agent tends to stick to soil and also to settle to the bottom of surface water, making the chemical agent less available to organisms.

Persistence of Pesticides in Soil

How persistent are pesticides in soil? Persistence is measured as the time it takes for half of the initial amount of a pesticide to break down. Thus if a pesticide's half-life is 30 days, half will be left after 30 days, one-quarter after 60 days, one-eighth after 90 days, and so on. It might seem that a short half-life would mean a pesticide would not have a chance to move far in the environment. However, if the pesticide is soluble in water and the conditions are favorable, it can move rapidly through certain soils. As the residual pesticide moves away from the surface, it also moves away from agents that can degrade it, such as sunlight and bacteria. As the pesticide moves deeper into the soil, it degrades more slowly and thus has a chance to enter groundwater.

DID YOU KNOW?

It has been estimated that 65 percent of all agricultural land in Europe contains high enough concentrations of pesticides to warrant health concern.

TABLE 6-1 Pesticide Persistence in Soil

Low (half-life <30 days)	Moderate (half-life 30–100 days)	High (half-life >100 days)
aldicarb	aldrin	bromacil
captan	atrazine	chlordane
calapon	carbaryl	lindane
dicamba	carbofuran	paraquat
malathion	diazinon	picloram
methyl-parathion	endrin	TCA
oxamyl	heptachlor	trifluralin

PESTICIDES AND HUMAN HEALTH

Pesticides are effectively used to protect crops and livestock from being eaten, sickened, or damaged by pest organisms. Nonetheless, pesticides can harm humans, other organisms, and the environment. For example, many pesticides are toxic to people if ingested, inhaled, or absorbed through the skin. Such pesticides can cause illness, disease, or even death.

Farmworkers who mix and apply pesticides are at particular risk from exposure through inhalation or skin contact. The World Health Organization (WHO) estimates that pesticides cause 3 million acute poisonings among agricultural workers worldwide each year. Experts estimate that the number of agricultural laborers poisoned by pesticides in the United States may range from 27,000 to 300,000 cases per year. Research

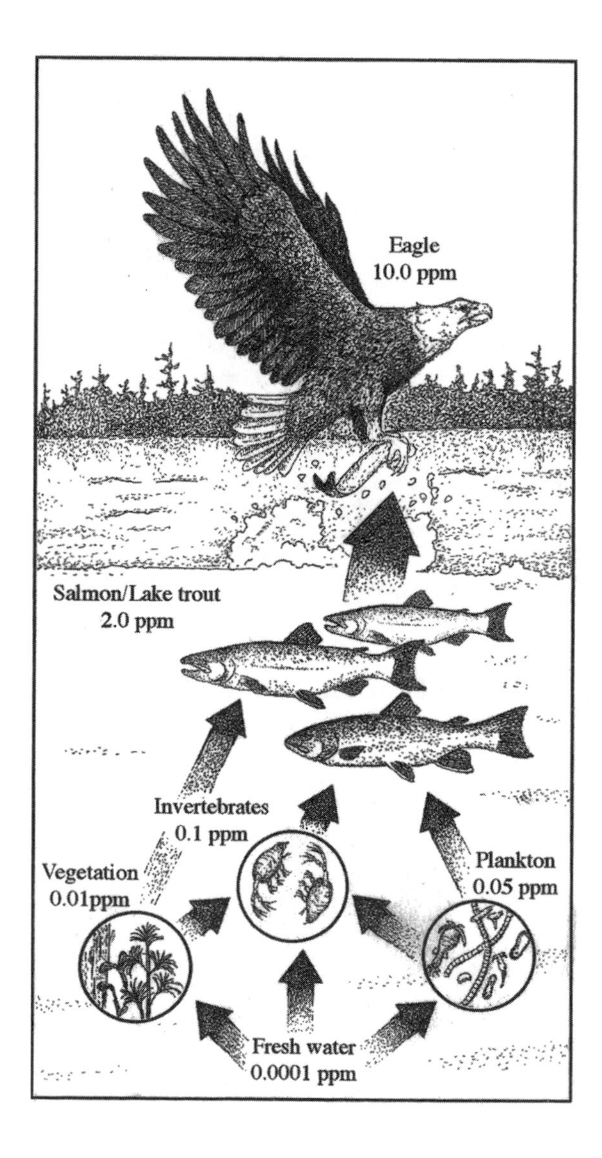

FIGURE 6-6 • Pesticides can affect organisms in aquatic food chains. Biological magnification is the increase of contaminants such as pesticides in organisms as a result of their consumption within the food chain. Notice the changes in the pesticide DDT (parts per million), as the concentration of the chemical moves through the food chain at different trophic levels. (Courtesy of Michael Kamrin, Michigan State University, *Ecotoxicology for the Citizen*)

has indicated that European farmers may be at increased risks of certain cancers, including malignant lymphoma, leukemia, multiple myeloma, and cancers of the testicles, gastrointestinal tract, lung, and brain.

Controlling and Regulated the Use of Pesticides

Federal Insecticide, Fungicide, and Rodenticide Act The Federal Insecticide, Fungicide, and Rodenticide Act (FIFRA) is a law that provides for federal control of pesticide distribution, sale, and use. The law is administered by the U.S. Environmental Protection Agency (EPA).

The EPA studies the consequences of pesticide use and requires users (farmers, utility companies, and others) to register when purchasing pesticides. All pesticides used in the United States must be registered (licensed) by the EPA. Registration includes approval by the EPA of the pesticide's label, which must give detailed instructions for its safe use.

The EPA classifies each pesticide to be used for general use, restricted use, or both. General use pesticides may be applied by anyone, but restricted use pesticides may be applied only by certified applicators or persons working under the direct supervision of a certified applicator. Registration assures that pesticides will be properly labeled and that if used in accordance with specifications will not cause unreasonable harm to the environment.

Vocabulary

Alien species A species that is not native to an area but has been introduced to the area by human activity.

Biome A large ecosystem that is characterized by vegetation, climate, and all living species in it.

Carrying capacity The maximum number of individuals of a species that can be supported in a given area.

Ecological succession A series of stages in which a population of organisms living in a biological community reaches its final stage.

Regeneration The growing again of vegetation on land.

Salinity The proportion of sodium chloride or other salts that are present in a given amount of soil or water.

Volatile Having the potential to evaporate into a gaseous state rapidly.

Activities for Students

1. Many places that are currently being deforested are in developing countries. What responsibility do developed countries have to support responsible use of forests in these areas?
2. Visit the World Wildlife Fund Website and view ways that they encourage sustainable forest management. How can you support and encourage others to support sustainable products from forests?
3. Many farmworkers are temporary or migrant workers. Create an informational poster to let farmworkers know about risks to their health presented by pesticides and how they can protect themselves.
4. Write a letter to a local representative explaining the problems created by environmental destruction of land and why he or she should work to protect land in your area.

Books and Other Reading Materials

Brown, Katrina, and David W. Pearce, eds. *The Causes of Tropical Deforestation: The Economic and Statistical Analysis of Factors Giving Rise to the Loss of the Tropical Forests*. Vancouver: University of British Columbia Press, 1994.

DeStefano, Susan. *Chico Mendes: Fight for the Forest*. Frederick, Md.: Twenty-First Century Books, 1992.

Gash, J.H.C., et al., eds. *Amazonian Deforestation and Climate,* New York: John Wiley & Sons, 1996.

Gradwohl, J., and R. Greenberg. *Saving the Tropical Forests.* Washington, D.C.: Island Press, 1988.

Hecht, S., and A. Cockburn. *The Fate of the Forest: Developers, Destroyers, and Defenders of the Amazon.* London: Verso, 1989.

Myers, Norman, consulting ed. *Rainforests*. Emmaus, PA: Rodale Press; New York: distributed in the book trade by St. Martin's Press, 1993.

Olson, Richard K., ed. *Integrating Sustainable Agriculture, Ecology, and Environmental Policy*. Binghamton, N.Y.: Food Products Press, 1992.

Schnoor, J.L., ed. *Fate of Pesticides and Chemicals in the Environment*. Environmental Science and Technology. New York: John Wiley & Sons, 1991.

Soule, Judith D., and Jon K. Piper. *Farming in Nature's Image: An Ecological Approach to Agriculture*. Washington, D.C.: Island Press, 1992.

Steen, H.K., and R.P. Tucker, eds. *Changing Tropical Forests: Historical Perspectives on Today's Challenges in Central and South America*. Durham, N.C.: Forest History Society, 1992.

Wargo, John. *Our Children's Toxic Legacy: How Science and Law Fail to Protect Us from Pesticides*. New Haven, Conn.: Yale University Press, 1998.

Websites

Greenpeace International, Forests, http:www.greenpeace.org/~forests

Pesticides in the Atmosphere, http://www.p510dcascr.wr.usgs.gov/pnsp/atmos/ atmos_4.html

Society of American Foresters, http://www.safnet.org

Toxics and Pesticides, http://www.epa.gov/oppfead1/work_saf/ (The Worker Protection Standard, or WPS, is intended to reduce the risk of pesticide poisonings and injuries among agricultural workers.)

U.S. Forest Service, http://www.fs.fed.us

World Wildlife Fund (Worldwide Fund for Nature) Forests for Life Campaign, http://www.panda.org/forests4life

CHAPTER 7

Decline of Wild Species and Habitats

Many environmentalists confirm that by the year of 2100 between 20 and 50 percent of all wildlife species evident on Earth since 1900 will become extinct. One environmental study reported that as many as 100 species disappear from Earth each day. What are some of the causes?

ENDANGERED AND THREATENED SPECIES

Wildlife species can become extinct because of climate change and other natural disasters such as droughts, landslides, earthquakes, and volcanic eruptions. Mass extinctions—catastrophic events—have also occurred periodically throughout Earth's history, destroying species and their habitats, however, human activities, such as overhunting and habitat destruction, can also place species at risk.

Species that are at the risk of extinction include endangered species and threatened species. An endangered species' total population is one in immediate danger of extinction. Some endangered species include the jaguar, white whale, Wyoming toad, and Atlantic salmon.

A threatened species is a species whose population is larger and more stable than that of a species recognized as endangered. But threatened species have the potential of becoming an endangered species. Examples of threatened species include the desert tortoise, grizzly bear, and Canada lynx. In 2001, there were 514 species of animal that were endangered or threatened and 740 species of plants listed as endangered or threatened in the United States.

DID YOU KNOW?

In the 1900s, 500,000 tigers lived in India. Presently there are less than 10,000 tigers living throughout the world.

Dodo Bird

The dodo bird (*Raphus cucullatus*), was a native bird of the island of Mauritius in the Indian Ocean. The bird was driven to extinction around the 1680s. In the 1500s, trading ships landed on the island and sailors hunted the dodos for food. Later non-native animals such as rats, pigs, and monkeys were introduced to the island; the animals preyed on the dodos and ate their eggs. The birds' habitat was also being destroyed as the island's forests were cut down and the land was converted to farming plantations. The combination of overhunting, destruction of habitats, and introduction of alien animals took its toll on the dodo population. Within 200 years of the arrival of humans on Mauritius, the dodo became extinct.

CAUSES OF EXTINCTION

Habitat Loss

Each species is adapted to life in a particular habitat. The habitat of an organism provides the

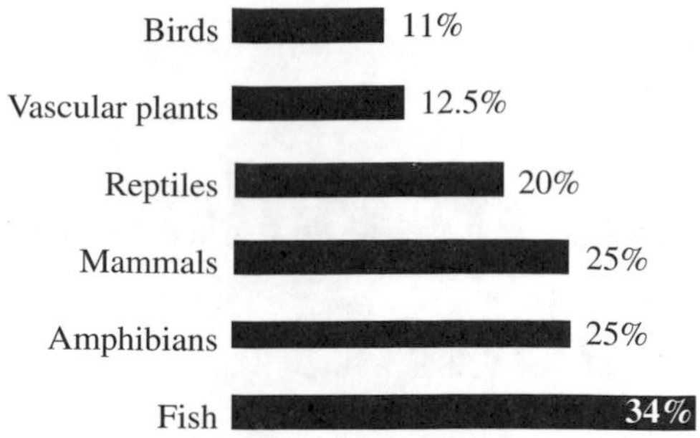

Group	Percentage
Birds	11%
Vascular plants	12.5%
Reptiles	20%
Mammals	25%
Amphibians	25%
Fish	34%

FIGURE 7-1 • Wildlife Under Pressure: Percentage of Species Threatened, 1996

FIGURE 7-2 • The dodo bird was a flightless bird that became extinct. It lived on many islands in the Indian Ocean.

FIGURE 7-3 • In 1914, the last known passenger pigeon, a female named Martha, died in captivity at the Cincinnati Zoo.

organism with all the biotic factors and abiotic factors such as food, water, shelter, proper temperature, and mates that are needed to sustain life and to ensure the survival of its species. Thus destruction or loss of habitat threatens the species' survival if organisms are unable to *adapt* to such changes or move to a new habitat. Habitat loss accounts for 85 percent of all species that are threatened with extinction.

There are natural occurrences that can result in habitat loss. These natural disasters include climate change and such events as floods, volcanic eruptions, and forest fires. Human activities also destroy or alter habitats. These human activities include the

- release of pollutants into air, water, and soil,
- diversion of water from rivers and streams to other areas for drinking water and crop irrigation,
- impact of deforestation, particularly in rainforests,
- drainage of water from marshes and mangroves for home sites and for aquaculture,
- overfishing and overhunting activities,
- clearing of land to develop roads, farmland, grazing pastures, housing developments, or shopping malls, and
- degradation of coral reefs by overfishing and pollution.

TABLE 7-1 Total Endangered Species in the United States and Worldwide

	Endangered		Threatened		Total
Group	U.S.	Foreign	U.S.	Foreign	Species
Mammals	61	251	8	16	336
Birds	75	178	15	6	274
Reptiles	14	65	21	14	114
Amphibians	9	8	8	1	26
Fish	69	11	41	0	121
Clams	61	2	8	0	71
Snails	18	1	10	0	29
Insects	28	4	9	0	41
Arachnids	5	0	0	0	5
Crustaceans	17	0	3	0	20
Animal subtotal	357	520	123	37	1,037
Flowering plants	540	1	132	0	673
Conifers	2	0	1	2	5
Ferns and others	26	0	2	0	28
Plant subtotal	568	1	135	2	706
Grand total	925	521	258	39	1,743

Source: United States Fish and Wildlife Service, 1998.

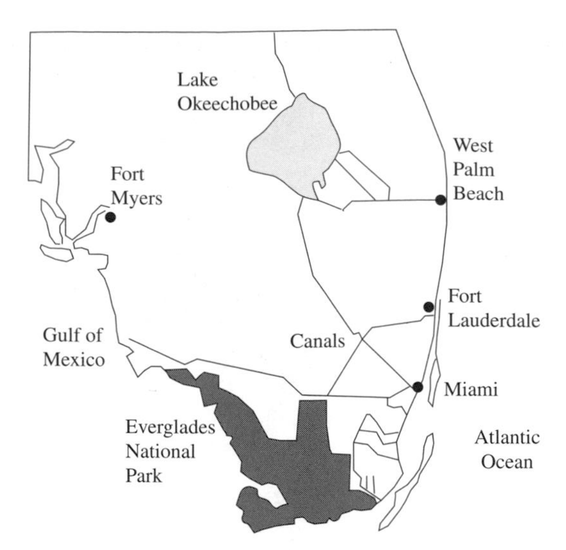

FIGURE 7-4 • Everglades National Park The diverse ecosystems of the Everglades provide habitat for abundant wildlife, including rare and colorful birds, manatees, turtles, snakes, alligators, the Florida panther, and crocodiles. However, a great deal of the water that reaches the Everglades is polluted with phosphates from agricultural and household runoff. Many species are disappearing, including the Florida panther, the wood stork, and the Cape Sable sparrow.

Habitats can be destroyed completely or broken up. When habitats are broken up, they are fragmented into "islands," where the organism populations of species are crowded into smaller areas. Island populations lose contact with other members of their species, thereby reducing *genetic* diversity and making the species less adaptable to environmental changes. Small populations are particularly vulnerable to extinction; for some species, the fragmented habitats become too small to support a viable population.

Poaching

The illegal hunting or trapping of wildlife in a no-hunting area or preserve or at restricted times of the year is called poaching. Poaching has caused animals to be placed on the threatened or endangered list. The greatest threat to both the Bengal and Siberian tigers, for instance, is the poaching of the animals to obtain their pelts and body parts such as the tail, teeth, bones, and whiskers. Poaching also threatens gorillas in areas where the animals are hunted for food or for body parts used to make ornaments to sell and trade. The population of Asia's two-horned rhinoceros has been reduced to about 300 animals because of poaching. Poachers sell the animal horns for medicinal purposes.

Alien Species

An alien species is any type of organism that is present in an *ecosystem* as a result of being introduced to that ecosystem through accidental or intentional means. Because they are not native to an area, alien species are sometimes referred to as exotic species, foreign species, or introduced species. Alien species are of concern to environmental scientists because they often disrupt the natural balance of the ecosystem to which they are introduced. One example of an alien species are the zebra mussels. The zebra mussels inhabit the waters of the Great Lakes but are not native to the area. Scientists reported that the zebra mussels were accidentally discharged from the ballasts of cargo ships from Asia that traveled to the ports on the Great Lakes. Over time, the zebra mussels have increased in population, causing problems such as the plugging up of freshwater intake pipes.

Alien species, such as the zebra mussels, often disrupt an ecosystem because they do not have natural predators in their new environments. This permits the population of the new species to grow virtually unchecked. The growing population may in turn then outcompete native species for such essential resources as food, water, and living space. When this occurs, the native species often abandons its habitat or begins to die off, sometimes placing it at risk of extinction.

DID YOU KNOW?

The kudzu is a vine that grows throughout the eastern United States. It is an alien species that was imported from Asia to help farmers control soil erosion on slopes. The plant grows so fast, however, that it is impossible to control its growth in many areas.

Overfishing

Overfishing is the process of continuing to fish in an area until all the fish are harvested. The breeding grounds become almost depleted or destroyed. According to the UN Food and Agriculture Organization (FAO), 69 percent of the world's fisheries are currently overexploited. Since the early 1950s, the demand for fish for human consumption has increased; to meet this demand, many fishing industries have invested

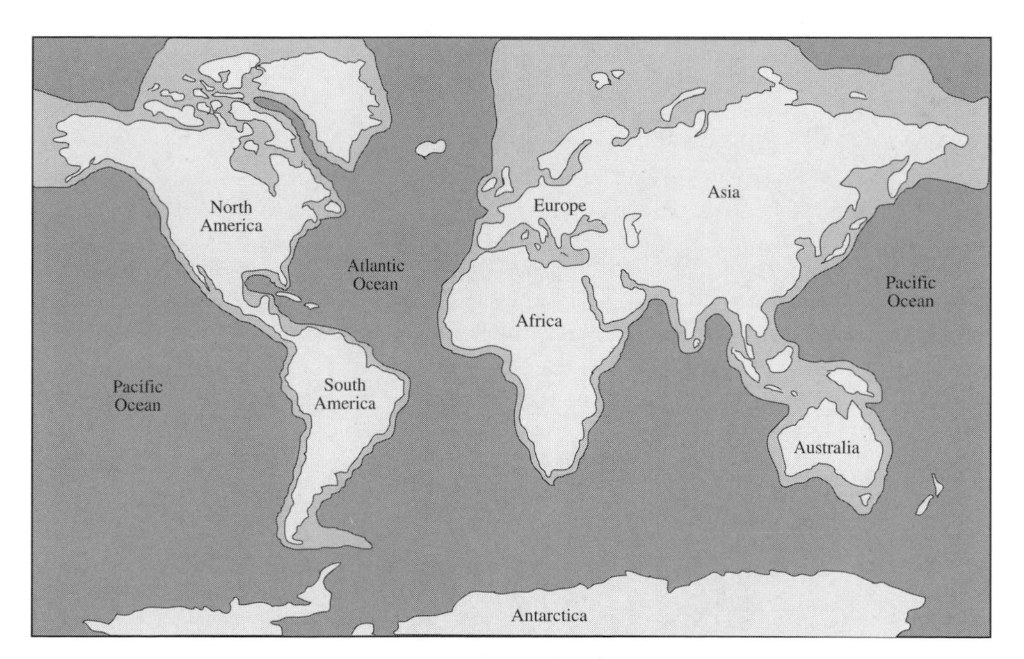

FIGURE 7-5 • The Continental Shelves of the World The continental shelves are the most productive area of the ocean with respect to commercial fishing. However, fish stocks have been declining in the waters of the continental shelves throughout the world.

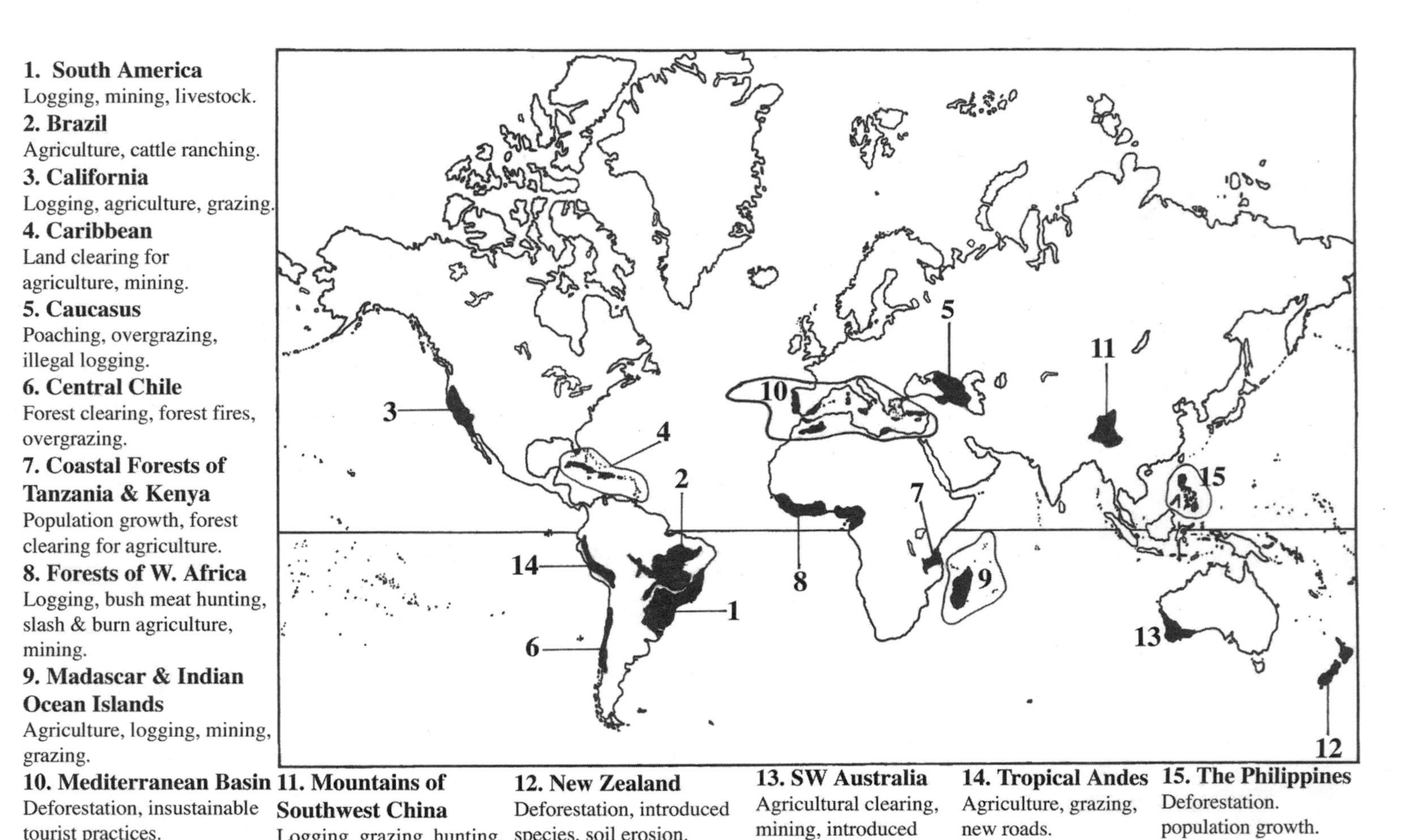

FIGURE 7-6 • Fading Habitats Some locations in the world are losing wildlife habitats due to poaching, population growth, agricultural and ranching activities, and logging.

in trawler fleets and ocean-going, factory-type processing vessels that can travel to fishing grounds all over the ocean. These vessels, equipped with sophisticated, on-board electronic tools and satellite data receivers can track large schools of fish day and night. Harvesting fish has also improved with the use of long drift nets, longliners, and purse seines that can catch thousands of kilograms (pounds) of fish in a brief period of time.

Overfishing is more severe in some regions than in others. One of the major areas in which fish stocks have been declining is the North Atlantic Ocean. Thousands of fishers in the United States and Canada have lost their jobs because of severe declines in cod, flounder, and haddock populations off the coasts of New England and Canada. In the northwestern section of the Pacific Ocean, 100 percent of coastal Asian fish stocks have been exploited. There have also been severe declines in salmon stock in the Pacific Ocean and declines in groupers in the Gulf of Mexico. More than 30 percent of the 280 species of fish in the United States' waters are in trouble. To add to the problem, many marine food webs will be destroyed if the rate of declining fish stocks continues. Many lakes, as well as the North Sea, Adriatic Sea, Black Sea, and Baltic Sea, have also suffered from overfishing. Another factor of decline includes the loss of "seafood breeding" estuaries and other spawning grounds of commerical fishing sites.

FIGURE 7-7 • The humpback (top), right whale (middle) and bowhead (bottom) are part of a group of marine mammals of the order *Cetacea* that have flippers and tails with horizontal flukes. Many types of whales are now threatened species or endangered species such as the humpback and bowhead.

Overhunting of Whales

Human beings have hunted whales for millennia, but commercial whaling in the nineteenth and twentieth centuries brought several populations of whales to the brink of extinction through overfishing. Some types of whales are now a threatened or endangered species. The most endangered whale is the humpback, known for its songs and acrobatics. Also endangered are the blue, bowhead, finback, right, sei, and sperm whales.

BIODIVERSITY

Environmentalists are concerned that as the human population continues growing and altering environments to meet its needs, more organisms will become extinct. This will decrease Earth's biodiversity. What is biodiversity, and why is it important?

Biodiversity is a term used to refer to the variety of plants, animals, fungi, protists, and bacteria living on Earth. Currently more than 1.5 million species of living organisms have been identified by scientists. About 75 percent of these known species are insects. Many scientists believe that the number of organisms that may inhabit Earth may be as much as ten times greater than the number of known species. These unknown species are believed to exist mostly in the regions of Earth that have not yet been thoroughly explored and identified by scientists. These remote places include areas in the tropical rainforests, coral reefs, and the deep ocean.

Value of Biodiversity

Maintaining Earth's biodiversity is a major goal of environmental science. Biodiversity is essential to ecosystem health. Because of the interactions that occur between Earth's organisms and the physical environment, the health of ecosystems is essential to maintaining the health of the biosphere. For example, organisms and the processes that keep them alive are essential to maintaining a proper balance of atmospheric gases such as oxygen, carbon dioxide, and nitrogen. Earth's water cycle also is largely dependent upon organisms. Any breakdown in the interaction of organisms and their physical environment can disrupt the health of the biosphere.

Many of the products humans rely on to maintain life or to make their lives more comfortable are also derived directly or indirectly from Earth's living things. For example, many life-saving medications have been derived from plants, particularly those in the rainforests. The drugs colchinine, taxol, and vinbalstine, which are used to prevent or treat different types of cancer, are derived from the autumn crocus, Pacific yew, and rosy periwinkle, respectively. The antibiotic penicillin

TABLE 7-2 Biodiversity of Known Species

	Number of species	
Animals	1,030,000	72.9%
Plants	250,000	17.7%
Fungi	69,000	4.9%
Protists	58,000	4.2%
Monerans	4,800	0.3%

originated from the penicillin mold, a fungus. Today the drug can be made synthetically. Quinine is an antimalarial drug derived from the yellow cinchona tree; the drug known as L-dopa, which is used in the treatment of Parkinson's disease, comes from the velvet bean plant.

Scientists point to the use of plants in making medicines as one of the reasons why preserving biodiversity is so important. Another reason is that many species that may contain similar medicinal substances have not yet been identified. Both of these facts emphasize why it is necessary to preserve Earth's ecosystems. Other products humans derive from organisms include a variety of foods; food products such as spices, wood, paper, dyes; and fibers used in the manufacture of clothing and other products.

Refer to Volume II for more information about the benefits of rainforest medicines.

TREATIES, LAWS, AND LISTS TO PROTECT BIODIVERSITY

Biodiversity Treaty

The Biodiversity Treaty is an international agreement developed to protect Earth's biodiversity. The treaty resulted from the UN Earth Summit held in Rio de Janeiro, Brazil, in 1992, where it was signed by more than 150 nations. The Biodiversity Treaty provides for the establishment of a worldwide inventory of threatened and endangered species and for the cooperation among nations to protect such species. The treaty also provides financial assistance from wealthy developed nations to less wealthy developing nations to help protect potentially valuable species.

Convention on International Trade in Endangered Species of Wild Fauna and Flora

Another environmental treaty that protects animal and plant species is the Convention on International Trade in Endangered Species of Wild Fauna and Flora (CITES). The 1973 international treaty protects more than 600 species of animals and plants, including the gorilla and rhinoceros. By the early 1990s, some success had been achieved in prohibiting trade in rhinoceros horn, elephant ivory, and endangered orchids. In many countries, however, a lack of local law enforcement, the willingness of some individuals to trade in endangered species, and the activities of poachers and traders put the future of many species in jeopardy despite the existence of legal protections.

Endangered Species Act

In the United States, the Endangered Species Act (ESA) was passed by Congress in 1973. The Endangered Species Act requires the conservation of threatened and endangered species and the ecosystems upon which they depend. The act also discourages the *exploitation* of endangered species in other countries by banning the importation or trade of any endangered species or any product made from such species.

The U.S. Fish and Wildlife Service (FWS) and the National Marine Fisheries Service are the two federal agencies in charge of managing the Endangered Species Act. Their responsibilities include working with private landowners, citizens, and organizations to

- conserve species,
- determine the species that need protection, and
- to restore listed species to a secure existence or recovery.

The two federal agencies cooperate with private landowners. Approximately 70 percent of all endangered and threatened species inhabit privately owned lands. Thus the cooperation and the involvement of the landowners are crucial to the management of at-risk species.

Endangered Species and Threatened Species List

A major FWS function is to identify and recover endangered species. The Fish and Wildlife Service leads the federal effort to protect and restore animals and plants that are in danger of extinction, both in the United States and worldwide. Using the best scientific information available, the agency identifies species that appear to be endangered or threatened. The species that meet the criteria of the Endangered Species Act are placed on the Department of Interior's official List of Endangered and Threatened Wildlife and Plants. For purposes of listing, the Fish and Wildlife Service defines endangered species as those at immediate risk of extinction that will not likely survive without direct human intervention. Threatened species are those that are abundant in parts of their range but are declining in total numbers. Thus they are at risk of extinction in the foreseeable future.

As of December 31, 2001, the Fish and Wildlife Service included 386 animal species and 595 plant species of the United States on its list of endangered species. The number of animal species from each group included 64 mammals, 78 birds, 14 reptiles, 11 amphibians, 71 fish, 62 clams, 21 snails, 35 insects, 12 arachnids, and 18 crustaceans. The number of U.S. plant species listed as endangered included more than 500 species of flowering plants, a few conifers, and a little more than 20 species of ferns or other plants. In addition, 128 animal species and 145 plant species are identified as threatened species.

Red List of Endangered Species

The Red List of Threatened and Endangered Species is a more complete survey of plant and animal species worldwide that are recognized as endangered or threatened. The Red List is maintained and published by the World Conservation Union (IUCN).

The Red List is similar but broader in scope than the U.S. Fish and Wildlife Service Endangered Species. The primary purpose of the Red

FIGURE 7-8 • The old man cactus is one of about 2,000 species of cacti, most of which are native to the southwestern United States, Mexico, Central America, and the southernmost countries of South America. Of the 2,000 known cacti species, approximately 20 are at risk of extinction, primarily because of over collecting by humans and damage caused by grazing animals.

TABLE 7-3 Endangered Tigers at a Glance

Bengal tiger (*Panthera tigris tigris*)	
Length	3 m (10 ft)
Weight	290 kg (650 lbs)
Estimated Population	4,000
Status	Endangered
Habitat	Mainland of Southeast Asia; India
Siberian tiger (*Panthera tigris altaica*)	
Length	4 m (13 ft)
Weight	300 kg (700 lbs)
Estimated Population	200
Status	Endangered
Habitat	Asia, as far north as Arctic Circle

TABLE 7-4 Endangered Cacti at a Glance

Common Name	Where Found
Tobusch fishhook cactus	Texas
Star cactus	Texas, Mexico
Fragrant prickly-apple	Florida
Nellie cory cactus	Texas
Key tree-cactus	Florida Keys
Arizona agave	Arizona

List is to increase global awareness of species that are at risk of going extinct. This way actions may be taken to prevent such occurrences.

Red Lists are published every few years, with separate lists being prepared for plants and animals. The lists focus on identifying species threatened with global extinction. Beginning in 1994, the IUCN created several classification categories for the organisms on its lists:

- Extinct
- Extinct in the wild
- Critically endangered
- Endangered
- Vulnerable

Extinct species are those for which all individuals of that species have ceased to exist. "Extinct in the wild" is used to classify organisms that are believed to now exist only in captivity (zoos, wildlife refuges, or captive breeding facilities), when cultivated (grown in nurseries or on croplands), or as a naturalized population. A naturalized population is one that survives outside its past range. "Critically endangered" is used for classifying organisms that are at high risk of extinction in the immediate future. This list is comparable to the FWS endangered species list. Endangered species are those facing a high risk of extinction in the wild, but with no timetable of extinction.

A species is deemed vulnerable when it is not critically endangered but poses a risk of extinction in the near future. The vulnerable species category is similar to the threatened species designation used by the Fish and Wildlife Service. The species that are identified as

FIGURE 7-9 • The African elephant is the largest land animal on Earth. African elephants have been brought to the brink of extinction by loss of their habitat and poaching for their ivory tusks. Poachers often disable elephants and leave them to die after severing their tusks.

FIGURE 7-10 • The mountain gorilla is an endangered species. The African Wildlife Foundation, an international organization established in 1961 to protect African wildlife from extinction, focuses on the African elephant, the mountain gorilla, and the black rhinoceros.

TABLE 7-5 Endangered Bats of the World at a Glance

Name	Where Found
Bumblebee bat Sanborn's lesser long-nosed bat	Thailand, Mexico, Central America, United States and Possessions
Mexican long-nosed bat	Mexico, Central America, United States and possessions
Rodriguez flying fox fruit bat	Rodriguez Island
Singapore roundleaf horseshoe bat	Malaysia
Gray bat; Hawaiian hoary bat; Indiana bat; Little Mariana fruit bat; Mariana fruit bat; Virginia big-eared bat; Ozark big-eared bat	United States and possessions

lower risk include population sizes that are not vulnerable, endangered, or critically endangered.

According to their world count, there are 11,096 species (5,485 animals, and 5,611 plants) listed as critically endangered, endangered, or vulnerable on the 2000 Red List. Of these, 1,939 are listed as critically endangered (925 animals, and 1,014 plants). The Red List also documents how human activities such as pollution and the fragmentation or loss of habitats have affected wildlife populations. They also study the introduction of alien species into an ecosystem.

According to the World Conservation Union, the main areas where mammals, birds, and plants (trees) seem to require the most conservation effort are in the Neotropics (Brazil, Colombia, Ecuador,

FIGURE 7-11 • The illustration shows a fishing bat. Bats live in almost all regions of the world except the Arctic. The main threats to bat populations are exposure to chemicals in the environment and habitat loss.

and Mexico), East Africa (Tanzania), and Southeast Asia (China, India, Indonesia, and Malaysia). For mammals alone, the situation appears most critical in Africa; whereas for the birds, Argentina and the Southeast Asian block of Myanmar, Vietnam, and Cambodia emerge as important areas.

PROGRAMS TO PROTECT WILDLIFE SPECIES AND HABITATS

To maintain Earth's biodiversity, scientists recommend that people learn as much as they can about the environment and use its resources wisely. This may help to prevent the extinction of many species in the future. In addition, several programs have been established to protect those species that have been identified as being on the verge of extinction. Such programs can be expensive to implement and are not always successful. However, until all of Earth's organisms are known and their roles in the environment understood, ignoring such efforts may prove to be even more costly to the future of humans and the environment. Wildlife refuges and captive propagation programs are just two of several programs that are designed to save wildlife species.

Wildlife Refuges

Wildlife refuges are areas of land and water that provide food, water, shelter, and space for wildlife. The wildlife refuges are set up to save some species from extinction, particularly those on the Endangered Species List. The refuges are usually maintained by a government or nonprofit organization for the preservation and protection of one or more wildlife species. Some refuges protect historical and archaeological sites.

U.S. Wildlife Refuge System

The U.S. Wildlife Refuge System consists of the world's largest and most diverse collection of lands specifically set aside for wildlife. The wildlife refuge system was initiated in 1903 when President Theodore Roosevelt designated Florida's Pelican Island as a *sanctuary* for pelicans and herons. Today 500 national wildlife refuges have been established, ranging in location from the Arctic Ocean to the South Pacific and from Maine to the Caribbean. Most refuges were created to protect the more than 800 species of migratory birds that travel along the four major north-south flyways. The United States has responsibilities under international treaties with Canada, Mexico, Japan, and Russia for migratory bird conservation.

As of 1992, the U.S. National Wildlife Refuge System administered by the U.S. Fish and Wildlife Service comprised some 503 areas. It coveris 36.5 million hectares (90 million acres) among the 50 states. Refuges in the United States range in size from Minnesota's Mille Lacs (less than an acre or hectare) to Alaska's Yukon Delta at about 9 million hectares (20 million acres). The vast majority of these lands are located in Alaska, with the rest spread across the United States and several U.S. territories.

Through the Partners for Wildlife Program, the Fish and Wildlife Service provides technical and financial assistance to private landowners. These landowners wish to restore wildlife habitat on their properties, primarily wetlands, *riparian* habitat, and native prairies. To date nearly 11,000 landowners have participated in Partners for Wildlife, restoring thousands of hectares of habitat.

Refer to Volume V for more information on saving wildlife and habitats.

International Reserves

There are many international refuges and preserves throughout the world. In many countries the United Nations has established protected areas called biosphere reserves. The biosphere reserves are designed to conserve the diversity of plants, animals, and microorganisms that make up the living biosphere. As of 1996 there were 337 biosphere reserves located in 85 countries with a total area of 219,891,487 hectares (547,000,000 acres). Forty-seven reserves are located in the United States. One of the largest biosphere reserves is the Serengeti National Park in Tanzania in southeastern Africa. The Serengeti National Park is located in the vast subtropical grassland that is home to more than 2 million wildebeest, 500,000 Thomson's gazelle, 250,000 zebra, and nearly 500 species of birds.

Refer to Volume I for more information about the Serengeti National Park.

Captive Propagation

Direct actions to save endangered species include captive propagation, or captive breeding. Captive propagation constitutes one method of combating species extinction. Captive propagation programs are

Tsavo National Park

Tsavo National Park in Kenya consists of the Tsavo Park East and Tsavo Park West. These national parks are refuges for many endangered elephants and other threatened species. The park was opened in 1948 and covers about 20,000 square kilometers (7,720 square miles). The parks are known as one of the world's leading biodiversity strongholds because they include a wide variety of vegetation and terrain. Over 600 species of birds have been recorded. Animals who live in the parks include the leopard, cheetah, buffalo, rhinoceros, elephant, crocodile, mongoose, giraffe, zebra, and lion. Elevations in the parks range from 160 to 2,000 meters (500 to 6,000 feet) above sea level. Included in Tsavo East is Yatta Plateau, the world's largest lava flow. The parks are popular sites for ecotourism and safaris because of their easy accessibility. The parks became famous through the notorious "Man Eaters of Tsavo" incident at the turn of the century, when lions were preying on the workers building the great Uganda Railway.

FIGURE 7-12 • The wildebeest live in Tsavo National Park in Kenya, a refuge for many endangered elephants and other threatened species.

especially useful for increasing the populations of rare and endangered species. The programs control the mating and breeding of captive animals and plants in zoos, aquariums, botanical gardens, and private research institutions.

Throughout the world, animal and plant species are disappearing largely because their habitats are being destroyed by human activities such as clear-cutting of forests, filling in of wetlands, or pollution. Animals are also hunted for food, medicines, and the use of their skins. This is troublesome to ecologists because when a population becomes too small, the genetic variation in the species is reduced. When this occurs, a species can have difficulty adapting to changes in its environment and may eventually become extinct.

Various environmental laws have been enacted to help protect endangered species; however, the only hope for saving some species from extinction is captive propagation. This is illustrated by the case of the California condor, North America's largest bird. The endangered giant birds' habitat once ranged from California to Florida. Human encroachment on the birds' habitat led to a severe decrease in the condor population. In 1980, scientists discovered that only 27 condors remained in the

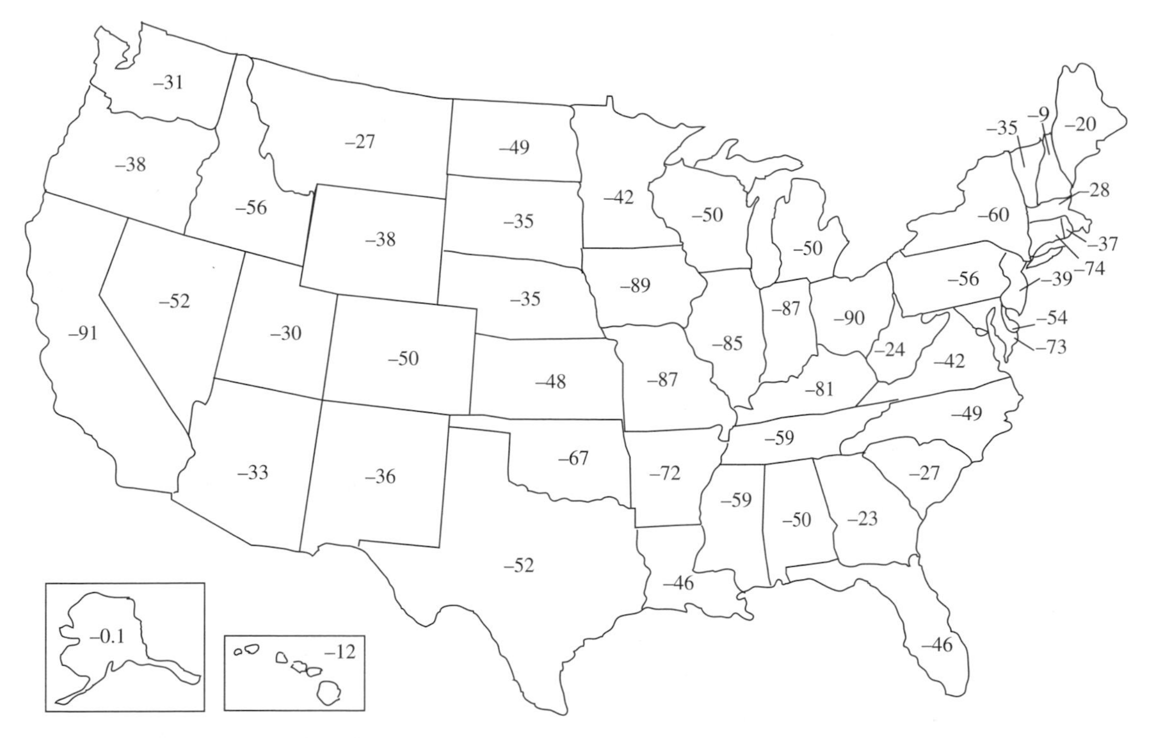

FIGURE 7-13 • Percentage of Wetland Acreage Lost between 1780 to 1980 All of the states have lost wetlands between 1780 and 1980. Twenty-two states lost at least 50 percent of their original wetlands. (Courtesy of United States Environmental Protection Agency, 1989)

FIGURE 7-14 • The California Condor usually nests in caves or on cliffs that have nearby trees for roosting and clear approaches for take-offs and landings.

wild. Five years later, the population plummeted to about nine condors. In 1987 the last nine wild condors were captured, leaving the animal effectively extinct in the wild. Today through captive propagation at the Los Angeles Zoo and the San Diego Zoo, scientists have increased the population of this rare bird. About 100 condors are still being held in captivity, but scientists have reintroduced about 13 condors into the wild.

Animal species that have benefited from being placed in captive propagation programs include the red wolf, the golden lion tamarin, the Arabian oryx, the Guam kingfisher, the Mongolian wild horse, and the Hawaiian goose (also called the nene). Overall, captive propagation programs have been quite successful; however, they are both time consuming and expensive. While some critically endangered species are rescued by such programs, hundreds of others move closer to extinction.

TABLE 7-6 California Condors at a Glance

Scientific Name	*Gymnogyps californianus*
Status	Endangered
Population Size	134 in the wild; 107 in captivity
Wingspan	2.5 m–3 m (8.5 ft–9.5 ft)
Mass	7.25 kg–10.8 kg (16 lbs–24 lbs)
Life span	40–50 years in wild; more in captivity

FIGURE 7-15 • The Oryx is an endangered species native to the Arabian Peninsula. Oryx are herd animals, living in groups of eight to twenty, composed of both sexes and all ages. Oryx live up to 20 years. Today the Oryx population is over 1,000 because of captive breeding techniques.

Vocabulary

Adapt To change to fit a new situation.

Ecosystem A system that includes all living organisms and the environment in which they live.

Exploitation Act of taking advantage of something in order to make money or a living or to gain other benefit.

Genetic Referring to the genes or biological inheritance.

Riparian Referring to the bank of a river.

Sanctuary A special area where wildlife is protected.

Activities for Students

1. Look at the U.S. Forest Service Website to discover where the closest wildlife refuge is to where you live. What kind of species of plants and animals can be found there?
2. Read the 1992 Biodiversity Treaty online. Write an updated treaty of your own, including resolutions about direct ways that governments can protect biodiversity.
3. Visit the World Conservation Monitoring Center website to learn about where wildlife refuges have been formed around the world. Create a world map that displays where refuges are located.
4. Visit your local zoo and compile a list of endangered species that are on display. Learn about any captive propagation efforts taking place there.

Books and Other Reading Materials

Berrill, Michael, and David Suzuki. *The Plundered Seas: Can the World's Fish Be Saved?* San Francisco: Sierra Club Books, 1997.

Chadwick, Douglas H., and Joel Sartore. *America's Endangered Species: The Company We Keep.* Washington, D.C.: National Geographic Society, 1998.

Ehrlich, Paul, and Anne Ehrlich. *Extinction: The Causes and Consequences of the Disappearance of Species.* New York: Random House, 1981.

Entering the Watershed: A New Approach to Save America's River Ecosystems. Bob Doppelt, Mary Scurlock, Chris Frissell, and James Karr. Sponsored by the Pacific Rivers Council. Washington, D.C.: Island Press. 1993.

Iudicello, Suzanne, et al. *Fish, Markets, and Fishermen: The Economies of Overfishing.* Washington, D.C.: Island Press, 1999.

Safina, Carl. *Song for the Blue Ocean: Encounters along the World's Coasts and beneath the Seas.* New York: Henry Holt & Company, 1998.

Wilson, Edward O. *The Diversity of Life.* New York: Belknap Press, 1992.

Websites

Convention on International Trade in Endangered Species of Wild Fauna and Flora (CITES), http://www.cites.org

Endangered Species List, U.S. Fish and Wildlife Service, http://www.fws.gov

Endangered Species & Wetlands Report, http://www.eswr.com

U.S. Fish & Wildlife Service, http://www.fws.gov/r9endspp/lsppinfo.html

U.S. Fish & Wildlife Service, *What You Need to Know . . . about the Endangered Species Act*, http://www.fws.gov/r9dia/public/esa.pdf

U.S. Fish & Wildlife Service Endangered Species Home Page, http://www.fws.gov/r9endspp/endspp.html

World Conservation Monitoring Center, http://www.wcmc.org

Appendix A: Environmental Timeline, 1620–2004

Environmentalists and activists appear in **boldface**.

1620 to 1860 Erosion becomes a major problem on many American farms. Fields are abandoned. Rivers and streams are filled with silt and mud. The publication of farm journals is initiated by early soil conservationists to improve farming methods.

1748 Jared Eliot, a minister and doctor of Killingsworth, Connecticut, writes the first American book on agriculture to improve crops and to conserve soil.

1824 Solomon and William Drown of Providence, Rhode Island, publish *Farmer's Guide* which discusses erosion and its causes and remedies. A year later, John Lorain, of the Philadelphia Agricultural Society, publishes a book devoted to the prevention of soil erosion in which he discusses methods such as using grass as an erosion-control crop.

1827 John James Audubon begins publication of *Birds of America*.

1830 George Catlin launches his great western painting crusade to document Native American peoples.

1845 Henry David Thoreau moves to Walden Pond to observe the fauna and flora of Concord, Massachusetts.

1847 U.S. Congressman **George Perkins Marsh** of Vermont delivers a speech calling attention to the destructive impact of human activity on the land.

1849 The U.S. Department of the Interior (DOI) is established.

1857 Frederick Law Olmsted develops the first city park: New York City's Central Park.

1859 British naturalist Charles Darwin publishes *The Origin of the Species by Means of Natural Selection*. In time the theory of evolution presented in the book becomes the most widely accepted theory of evolution.

1866 German biologist Ernst Haeckel introduces the term *ecology*.

1869 John Muir moves to the Yosemite Valley.

Geologist and explorer John Wesley Powell travels the Colorado River through the Grand Canyon.

1872 Yellowstone National Park is established as the first national park of the United States in Yellowstone, Wyoming.

U.S. legislation: Passage of the Mining Law permits individuals to purchase rights to mine public lands.

1876 The Appalachian Mountain Club is founded.

1879 The U.S. Geological Survey (USGS) is formed.

1882 The first hydroelectric plant opens on the Fox River in Wisconsin.

1883 Krakatoa, a small island of Indonesia, is virtually destroyed by a volcanic explosion.

1890 Denmark constructs the first windmill for use in generating electricity.

Sequoia National Park, Yosemite National Park, and General Grant National Park are established in California.

1891 U.S. legislation: Passage of Forest Reserve Act provides the basis for a system of national forests.

1892 John Muir, Robert Underwood Johnson, and William Colby are cofounders of the Sierra Club, in Muir's words, to "do something for wildness and make the mountains glad."

1893 The National Trust is founded in the United Kingdom. The group purchases land deemed of having natural beauty or considered a cultural landmark.

1895 Founding of the American Scenic and Historic Preservation Society.

1898 Cornell University establishes the first college program in forestry.

Gifford Pinchot becomes head of the U.S. Division of Forestry (now the U.S. Forest Service) and serves until 1910. Under President

Theodore Roosevelt, many of Pinchot's ideas became national policy. During his service, the national forests increase from 32 in 1898 to 149 in 1910, a total of 193 million acres.

1899 The River and Harbor Act bans pollution of all navigable waterways. Under the act, the building of any wharves, piers, jetties, and other structures is prohibited without congressional approval.

1900 U.S. legislation: Passage of Lacey Act makes it unlawful to transport illegally killed game animals across state boundaries.

1902 U.S. legislation: Passage of Reclamation Act establishes the Bureau of Reclamation.

1903 First federal U.S. wildlife refuge is established on Pelican Island in Florida.

1905 The National Audubon Society, named for wildlife artist John James Audubon, is founded.

1906 Yosemite Valley is incorporated into Yosemite National Park.

1907 International Association for the Prevention of Smoke is founded. The group's name later changes several times to reflect other concerns over causes of air pollution.

Gifford Pinchot is appointed the first chief of the U.S. Forest Service.

1908 The Grand Canyon is set aside as a national monument.

Chlorination is first used at U.S. water treatment plants.

President Theodore Roosevelt hosts the first Governors' Conference on Conservation.

1914 The last passenger pigeon, Martha, dies in the Cincinnati zoo.

1916 The National Park Service (NPS) is established.

1918 Hunting of migratory bird species is restricted through passage of the Migratory Bird Treaty Act. The act supports treaties between the United States and surrounding nations.

Save-the-Redwoods League is created.

1920 U.S. legislation: Passage of the Mineral Leasing Act regulates mining on federal lands.

1922 The Izaak Walton League is organized under the direction of **Will H. Dilg**.

1924 Environmentalist **Aldo Leopold** wins designation of Gila National Forest, New Mexico, as first extensive wilderness area.

Marjory Stoneman Douglas, of the *Miami Herald*, writes newspaper columns opposing the draining of the Florida Everglades.

Bryce Canyon National Park is established in Utah.

1925 The Geneva Protocol is signed by numerous countries as a means of stopping use of biological weapons.

1928 The Boulder Canyon Project (Hoover Dam) is authorized to provide irrigation, electric power, and a flood-control system for Arizona and Nevada communities.

1930 Chlorofluorocarbons (CFCs) are deemed safe for use in refrigerators and air conditioners.

1931 France builds and makes use of the first Darrieus aerogenerator to produce electricity from wind energy.

Addo Elephant National Park is established in the Eastern Cape region of South Africa to provide a protected habitat for African elephants.

1932 Hugh Bennett is given the opportunity to put his soil conservation ideas into practice to help reduce soil erosion. He becomes the director of the Soil Erosion Service (SES) created by the Department of Interior.

1933 The Tennessee Valley Authority (TVA) is formed.

The Civilian Conservation Corps (CCC) employs more than 2 million Americans in forestry, flood control, soil erosion, and beautification projects.

1934 The greatest drought in U.S. history continues. Portions of Texas, Oklahoma, Arkansas, and several other midwestern states are known as the "Dust Bowl."

U.S. legislation: Passage of Taylor Grazing Act regulates livestock grazing on federal lands.

1935 The Soil Conservation Service (SCS) is established.

The Wilderness Society is founded.

1936 The National Wildlife Federation (NWF) is formed.

1939 David Brower produces his first nature film for the Sierra Club, called *Sky Land Trails of the Kings*. In the same year, Brower, who is an excellent climber, completes his most famous ascent, Shiprock, a volcanic plug which rises 1,400 feet from the floor of the New Mexico desert.

1940 The U.S. Wildlife Service is established to protect fish and wildlife.

U.S. legislation: President Franklin Roosevelt signs the Bald Eagle Protection Act.

1945 The United Nations (UN) establishes the Food and Agriculture Organization (FAO).

1946 The International Whaling Commission (IWC) is formed to research whale populations.

The U.S. Bureau of Land Management (BLM) and the Atomic Energy Commission (AEC) are created.

1947 Marjory Stoneman Douglas publishes *The Everglades: River of Grass* and serves as a member of the committee that gets the Everglades designated a national park.

1948 The UN creates the International Union for the Conservation of Nature (IUCN) as a special environmental agency.

An air pollution incident in Donora, Pennsylvania, kills 20 people; 14,000 become ill.

U.S. legislation: Passage of Federal Water Pollution Control Law.

1949 Aldo Leopold's *A Sand County Almanac* is published posthumously.

1950 Oceanographer **Jacques Cousteau** purchases and transforms a former minesweeper, the *Calypso*, into a research vessel which he uses to increase awareness of the ocean environment.

1951 Tanzania begins its national park system with the establishment of the Serengeti National Park.

1952 Clean air legislation is enacted in Great Britain after air pollution–induced smog brings about the deaths of nearly 4,000 people.

David Brower becomes the first executive director of the Sierra Club.

1953 Radioactive iodine from atomic bomb testing is found in the thyroid glands of children living in Utah.

1955 U.S. legislation: Passage of the Air Pollution Control Act, the first federal legislation designed to control air pollution.

1956 U.S. legislation: Passage of the Water Pollution Control Act authorizes development of water-treatment plants.

1959 The Antarctic Treaty is signed to preserve natural resources of the continent.

1961 The African Wildlife Foundation (AWF) is established as an international organization to protect African wildlife.

1962 **Rachel Carson** publishes *Silent Spring*, a groundbreaking study of the dangers of DDT and other insecticides.

Hazel Wolf joins the National Audubon Society in Seattle, Washington, and plays a prominent role in local, national, and international environmental efforts during her lifetime.

1963 The Nuclear Test Ban Treaty between the United States and the Soviet Union stops atmospheric testing of nuclear weapons.

U.S. legislation: Passage of the first Clean Air Act (CAA) authorizes money for air pollution control efforts.

1964 **Hazel Henderson** organizes women in a local play park in New York City and starts a group called Citizens for Clean Air, the first environmental group, she believes, east of the Mississippi. She built Citizens for Clean Air from a very small group to a membership of 40,000. Two years later, 80 people died in New York City from air pollution–related causes during four days of atmospheric inversion.

U.S. legislation: Passage of the Wilderness Act creates the National Wilderness Preservation System.

1965 U.S. legislation: Passage of the Water Quality Act authorizes the federal government to set water standards in absence of state action.

1966 Eighty people in New York City die from air pollution–related causes.

1967 The *Torey Canyon* runs aground spilling 175 tons of crude oil off Cornwall, England.

Dian Fossey establishes the Karisoke Research Center in the Virunga Mountains, within the Parc National des Volcans in Rwanda to study endangered mountain gorillas.

The Environmental Defense Fund (EDF) is formed to lead an effort to save the osprey from DDT.

1968 U.S. legislation: Passage of the Wild and Scenic Rivers Act and the National Trails System Act identify areas of great scenic beauty for preservation and recreation.

Paul Ehrlich publishes *The Population Bomb*.

1969 Wildlife photographer Joy Adamson establishes the Elsa Wild Animal Appeal, an organization dedicated to the preservation and humane treatment of wild and captive animals.

Greenpeace is created.

Blowout of oil well in Santa Barbara, California, releases 2,700 tons of crude oil into the Pacific Ocean.

U.S. legislation: Passage of the National Environmental Policy Act (NEPA) requires all federal agencies to complete an environmental impact statement for any dam, highway, or other large construction project undertaken, regulated, or funded by the federal government.

The Friends of the Earth (FOE) is founded in the United States.

John Todd, **Nancy Jack Todd**, and Bill McLarney are the cofounders of the New Alchemy Institute in Cape Cod, Massachusetts. The institute begins to pioneer a new way of treating sewage and other wastes.

1970 **Denis Hayes** is the national coordinator of the first Earth Day, which is celebrated on April 22.

Construction of the Aswan High Dam on the Nile River in Egypt is completed.

U.S. legislation: Passage of an amended Clean Air Act (CAA) expands air pollution control.

The U.S. Environmental Protection Agency (EPA) is established.

1971 Canadian primatologist Biruté Galdikas begins her studies of orangutans through the Orangutan Research and Conservation Project in Borneo.

The United Nations Educational, Scientific and Cultural Organization (UNESCO) establishes the Man and the Biosphere Program, developing a global network of biosphere reserves.

1972 The Biological and Toxin Weapons Convention is adopted by 140 nations to stop the use of biological weapons.

The EPA phases out the use of DDT in the United States to protect several species of predatory birds. The ban builds on information obtained from Rachel Carson's 1962 book, *Silent Spring*.

U.S. legislation: Passage of the Water Pollution Control Act, the Coastal Zone Management Act (CZMA), and the Environmental Pesticide Control Act.

Oregon passes the first bottle-recycling law.

1973 Norwegian philosopher Arne Naess coins the term *deep ecology* to describe his belief that humans need to recognize natural things for their intrinsic value, rather than just for their value to humans.

The Convention on International Trade in Endangered Species of Wild Fauna and Flora (CITES) is signed by more than 80 nations. The Endangered Species Act of the United States also is enacted.

Congress approves construction of the 1,300-kilometer pipeline from Alaska's North Slope oil field to the Port of Valdez.

An Energy crisis in the United States arises from an Arab oil embargo.

A collision and resulting explosion between the *Corinthos* oil tanker and the *Edgar M. Queeny* releases 272,000 barrels of crude oil and other chemicals into the Delaware River near Marcus Hook, Pennsylvania.

1974 Scientists report their discovery of a hole in the ozone layer above Antarctica.

U.S. legislation: Passage of the Safe Drinking Water Act sets standards to protect the nation's drinking water. The EPA bans most uses for aldrin and dieldrin and disallows the production and importation of these chemicals into the United States.

1975 Unleaded gas goes on sale. New cars are equipped with antipollution catalytic converters.

The EPA bans use of asbestos insulation in new buildings.

Edward Abbey publishes *The Monkey Wrench Gang*, a novel detailing acts of ecotage as a means of protecting the environment.

1976 *Argo Merchant* runs aground releasing 25,000 tons of fuel into the Atlantic Ocean near Nantucket, Rhode Island.

National Academy of Sciences reports that CFC gases from spray cans are damaging the ozone layer.

U.S. legislation: Passage of the Resource Conservation and Recovery Act empowers the EPA to regulate the disposal and treatment of municipal solid and hazardous wastes. The Toxic Substances Control Act and the Resource Conservation and Recovery Act are enacted.

Fire aboard the *Hawaiian Patriot* releases nearly 100,000 tons of crude oil into the Pacific Ocean.

1977 The Green Belt Movement is begun by Kenyan conservationist Wangari Muta Maathai on World Environment Day.

Blowout of Ekofisk oil well releases 27,000 tons of crude oil into the North Sea.

Construction of the Alaska pipeline, the 1,300-kilometer pipeline that carries oil from

Alaska's North Slope oil field to the Port of Valdez, is completed at a cost of more than $8 billion.

U.S. legislation: Passage of the Surface Mining Control and Reclamation Act.

The Department of Energy (DOE) is created.

1978 The *Amoco Cadiz* tanker runs aground spilling 226,000 tons of oil into the ocean near Portsall, Brittany.

People living in the Love Canal community of New York are evacuated from the area to reduce their exposure to chemical wastes which have surfaced from a canal formerly used as a dump site.

Rainfall in Wheeling, West Virginia, is measured at a pH of 2, the most acidic rain yet recorded.

Aerosols with fluorocarbons are banned in the United States.

The EPA bans the use of asbestos in insulation, fireproofing, or decorative materials.

1979 British scientist **James E. Lovelock** publishes *Gaia: A New Look at Life on Earth.*

Collision of the *Atlantic Empress* and the *Aegean Captain* releases 370,000 tons of oil into the Caribbean Sea.

The Convention on Long-Range Transboundary Air Pollution (LRTAP) is signed by several European nations to limit sulfur dioxide emissions which cause acid rain problems in other countries.

The Three Mile Island Nuclear Power Plant in Pennsylvania experiences near-meltdown.

The EPA begins a program to assist states in removing flaking asbestos insulation from pipes and ceilings in school buildings throughout the United States.

The EPA bans the marketing of herbicide Agent Orange in the United States.

1980 Debt-for-nature swap idea is proposed by Thomas E. Lovejoy: nations could convert debt to cash which would then be used to purchase parcels of tropical rain forest to be managed by local conservation groups.

Global Report to the President addresses world trends in population growth, natural resource use, and the environment by the end of the century, and calls for international cooperation in solving problems.

U.S. legislation: Passage of the Comprehensive Environmental Response, Compensation, and Liability Act (Superfund) and the Low Level Radioactive Waste Policy Act.

1981 Earth First!, a radical environmental action group that resorts to ecotage to gain its objectives, formed.

Lois Gibbs founds the Citizens' Clearinghouse for Hazardous Wastes, later named the Center for Health, Environment, and Justice (CHEJ).

1982 U.S. legislation: Passage of the Nuclear Waste Policy Act.

1983 A film of **Randy Hayes**, *The Four Corners, a National Sacrifice Area*, wins the 1983 Student Academy Award for the best documentary. The film documents the tragic effects of uranium and coal mining on Hopi and Navajo Indian lands in the American Southwest.

The residents of Times Beach, Missouri, are ordered to evacuate their community. Investigations of Times Beach in the 1980s disclosed the fact that oil contaminated with dioxin, a highly toxic substance, had been used to treat the town's streets.

Cathrine Sneed founds and acts as director of the Garden Project in San Francisco. The Garden Project, a horticulture class for inmates of the San Francisco County Jail, uses organic gardening as a metaphor for life change. The U.S. Department of Agriculture calls the project "one of the most innovative and successful community-based crime prevention programs in the country."

1984 Toxic gases released from the Union Carbide chemical manufacturing plant in Bhopal, India kill an estimated 3,000 people and injure thousands of others.

The Jane Goodall Institute (JGI) is founded.

The British tanker *Alvenus* spills 0.8 million gallons of oil into the Gulf of Mexico.

U.S. legislation: Passage of the Hazardous and Solid Waste Amendments.

1985 Concerned Citizens of South Central Los Angeles becomes one of the first African American environmental groups in the United States. **Julia Tate** serves as the executive director. The organization's goal is to provide a better quality of life for the residents of this Los Angeles community. **Maria Perez**, **Nevada Dove**, and **Fabiola Tostado** later join the group and are known as the Toxic Crusaders.

Huey D. Johnson becomes the founder and president of the Resource Renewal Institute

(RRI), a nonprofit organization located in California. Johnson suggests that green plans is the path countries should take to respond to environmental decline. Green plans treat the environment as it really exists—a single, interconnected ecosystem that can be safeguarded for future generations only through a systemic, long-range plan of action.

Scientists of the British Antarctica Survey discover the ozone hole. The hole, which appears during the Antarctic spring, is caused by the chlorine from CFCs.

Juana Gutiérrez becomes president and founder of Mothers of East Los Angeles, Santa Isabel Chapter (Madres del Este de Los Angeles—Santa Isabel) (MELASI) whose mission is to fight against toxic dumps and incinerators and also to take a proactive approach to community improvement.

Primatologist Dian Fossey is discovered murdered in her cabin at the Karosoke Research Center she founded. Her death is attributed to poachers.

While protesting nuclear testing being conducted by France in the Pacific Ocean, the *Rainbow Warrior* (a boat owned by Greenpeace) is sunk in a New Zealand harbor by agents of the French government.

U.S. legislation: Passage of the Food Security Act.

1986 Tons of toxic chemicals stored in a warehouse owned by the Sandoz pharmaceutical company are released into the Rhine River near Basel, Switzerland. The effects of the spill are experienced in Switzerland, France, Germany, and Luxembourg.

An explosion destroys a nuclear power plant in Chernobyl, Ukraine, immediately killing more than 30 people and leading to the permanent evacuations of more than 100,000 others.

Bovine spongiform encephalopathy (BSE), a neurodegenerative illness of cattle, also known as mad cow disease, comes to the attention of the scientific community when it appears in cattle in the United Kingdom.

U.S. legislation: Passage of the Emergency Response and Community Right-to-Know Act and the Superfund Amendments and Reauthorization Act (SARA).

1987 The Montreal Protocol, an international treaty that proposes to cut in half the production and use of CFCs, is approved by more than 30 nations.

The world's fourth largest lake, the Aral Sea of Asia, is divided in two as a result of the diversion of water from its feeder streams, the Syr Darya and Amu Darya rivers.

The *Mobro*, a garbage barge from Long Island, New York, travels 9,600 kilometers in search of a place to offload the garbage it carries.

1988 Use of ruminant proteins in the preparation of cattle feed is banned in the United Kingdom to prevent outbreaks of BSE.

Global temperatures reach their highest levels in 130 years.

The Ocean Dumping Ban legislates international dumping of wastes in the ocean.

EPA studies report that indoor air can be 100 times as polluted as outdoor air. Radon is found to be widespread in U.S. homes.

Beaches on the east coast of the United States are closed because of contamination by medical waste washed onshore.

The United States experiences its worst drought in 50 years.

Plastic ring six-pack holders are required to be made degradable.

U.S. legislation: Passage of the Plastic Pollution Research and Control Act bans ocean dumping of plastic materials.

1989 The United Kingdom bans the use of cattle brains, spinal cords, tonsils, thymuses, spleens, and intestines in foods intended for human consumption as a means of preventing further outbreaks of Creutzfeldt-Jakob disease (CJD), the human version of mad cow disease, in humans.

Fire aboard the *Kharg 5* releases 75,000 tons of oil into the sea surrounding the Canary Islands.

The Montreal Protocol treaty is updated and amended.

The New York Department of Environmental Conservation reports that 25 percent of the lakes and ponds in the Adirondacks are too acidic to support fish.

The *Exxon Valdez* runs aground on Prince William Sound, Alaska, spilling 11 million gallons of oil into one of the world's most fragile ecosystems.

1990 Ocean Robbins, age 16, and **Ryan Eliason**, 18, are the cofounders of YES!, or Youth for Environmental Sanity. The goal of YES! is to educate, inspire, and empower young people to take positive action for healthy people and a healthy planet. Robbins served as director for five years and is now the organization's president. As of 2000, the program has reached 600,000 students in 1,200 schools in 43 states through full school assemblies.

UN report forecasts a world temperature increase of 2°F within 35 years as a result of greenhouse gas emissions.

U.S. legislation: Passage of the Clean Air Act amendments including requirements to control the emission of sulfur dioxide and nitrogen oxides.

1991 The Gulf War concludes with hundreds of oil wells in Kuwait being set afire by Iraqi troops, resulting in extensive air and water pollution problems.

The United States accepts an agreement on Antarctica which prohibits activities relating to mining, protects native species of flora and fauna, and limits tourism and marine pollution.

Eight scientists begin a two-year stay in Biosphere 2 in Arizona, a test center designed to provide a self-sustaining habitat modeling Earth's natural environments. The experiment, which is repeated in 1993, meets with much criticism and is deemed largely unsuccessful.

1992 UN Earth Summit is held in Rio de Janeiro, Brazil. Major resolutions resulting from the summit include the Rio Declaration on Environment and Development, Agenda 21, Biodiversity Convention, Statement of Forest Principles, and the Global Warming Convention, which is signed by more than 160 nations.

Severn Cullis-Suzuki speaks for six minutes to the delegates urging them to work hard on resolving global environmental issues. She received a standing ovation.

The Montreal Protocol is again amended with signatories agreeing to phase out CFC use by the year 2000.

1993 Sugar producers and U.S. government agree on a restoration plan for the Florida Everglades.

1994 *Dumping in Dixie: Class and Environmental Quality* is published by **Robert Bullard**. The book reports on five environmental justice campaigns in states ranging from Texas to West Virginia. Bullard emphasizes that African Americans are concerned about and do participate in environmental issues.

The California Desert Protection Act is passed.

Failure of a dike results in the release of 102,000 tons of oil into the Siberian tundra near Usink in northern Russia.

The Russian government calls for preventive measures to control the destruction of Lake Baikail.

The bald eagle is reclassified from an endangered species to a threatened species on the U.S. Endangered Species List.

An 8.5-million-gallon spill is discovered in Unocal's Guadalupe oil field in California.

1995 The U.S. Government reintroduces endangered wolves to Yellowstone Park.

1999 Scientists report that the human population of Earth now exceeds 6 billion people.

The peregrine falcon is removed from the U.S. Endangered Species List.

The *New Carissa* runs aground off the coast of Oregon, leaking some oil into Coos Bay. The tanker is later towed into the open ocean and sunk.

Beyond Globalization: Shaping a Sustainable Global Economy is published by Hazel Henderson.

Paul Hawken coauthors *Natural Capitalism, Creating the Next Industrial Revolution.*

Off the Map, an Expedition Deep into Imperialism, the Global Economy, and Other Earthly Whereabouts is published by **Chellis Glendinning**.

Twenty-three-year-old **Julia Butterfly Hill** comes down out of a 180-foot California redwood tree after living there for two years to prevent the destruction of the forest. A deal is made with the logging company to spare the tree as well as a three-acre buffer zone.

2000 Denis Hayes is the coordinator and **Mark Dubois** is the international coordinator of Earthday 2000.

Ralph Nader and **Winona LaDuke** run for U.S. president and vice president on the Green Party ticket.

In January 2000, Hazel Wolf passes away at the age of 101.

The Chernobyl nuclear power plant is scheduled to close in December.

Anthropologists for the Wildlife Conservation Society in New York announce that a type of large West African monkey is extinct, making it the first primate to vanish in the twenty-first century.

A study by National Park Trust, a privately funded land conservancy, finds that more than 90,000 acres within state parks in 32 states are threatened by commercial and residential development and increased traffic, among other things.

A bone-dry summer in north-central Texas breaks the Depression-era drought record when the Dallas area marks 59 days without rain. The arid streak with 100-degree daily highs breaks a record of 58 days set in the midst of the Dust Bowl in 1934 and tied in 1950. The Texas drought exceeded 1 billion dollars in agricultural losses.

Massachusetts announces that the state will spend $600,000 to determine whether petroleum pollution in largely African American city neighborhoods contributes to lupus, a potentially deadly immune disease. The research, to be conducted over three years, will target three areas of the city with unusually high levels of petroleum contamination.

Hybrid vehicle Toyota Prius is offered for sale in the United States.

The hole in the ozone layer over Antarctica has stretched over a populated city for the first time, after ballooning to a new record size. Previously, the hole had opened only over Antarctica and the surrounding ocean.

2001 An environmental group that successfully campaigned for the return of wolves to Yellowstone National Park wants the federal government to do the same in western Colorado and parts of Utah, southern Wyoming, northern New Mexico, and Arizona.

The UN Environment Program launches a campaign to save the world's great apes from extinction, asking for at least $1 million to get started.

The captain and crew of a tanker that spilled at least 185,000 gallons of diesel into the fragile marine environment of the Galapagos Islands have been arrested.

One hundred sixty-five countries approve the Kyoto rules aimed at halting global warming. The Kyoto Protocol requires industrial countries to scale back emissions of carbon dioxide and other greenhouse gases by an average of 5 percent from their 1990 levels by 2012. The United States, the world's biggest polluter rejects the pact.

The EPA reaches an agreement for the phaseout of a widely used pesticide, diazinon, because of potential health risks to children.

For the second time in three years, the average fuel economy of new passenger cars and light trucks sold in the United States dropped to its lowest level since 1980.

More and more Americans are breathing dirtier air, and larger U.S. cities such as Los Angeles and Atlanta remain among the worst for pollution.

In rural stretches of Alaska, global warming has thinned the Arctic pack ice, making travel dangerous for native hunters. Traces of industrial pollution from distant continents is showing up in the fat of Alaska's marine wildlife and in the breast milk of native mothers who eat a traditional diet including seal and walrus meat.

2002 A Congo volcano devastates a Congolese town burning everything in its path, creating a five-foot-high wall of cooling stone, and leaving a half million people homeless.

New research is conducted in the practice of killing sharks solely for their fins.

A report by the USGS shows the nation's waterways are awash in traces of chemicals used in beauty aids, medications, cleaners, and foods. Among the substances are caffeine, painkillers, insect repellent, perfumes, and nicotine. These substances largely escape regulation and defy municipal wastewater treatment.

A microbe is discovered to be a major cause of the destruction of beech trees in the northeastern United States.

A study discovers that, if fallen leaves are left in stagnant water, they can release toxic mercury, which eventually can accumulate in fish that live far downstream.

Scientists are experimenting with various sprays containing clay particles to kill toxic algae in seawater.

Meteorologists discover that the Mediterranean Sea receives air current pollutants from Europe, Asia, and North America.

Researchers report possible ways of blocking the deadly effects of anthrax.

2003 A new international treaty—The Protocol on Persistent Organic Pollutants (POPS) was ratified by 17 nations although the United States has not signed on. The treaty drafted by United Nations reduces and eliminates 16 toxic chemicals that are long-lived in the environment and travel globally. The new treaty, an extension of an earlier one signed in 2000, added four more organic persistent pollutants to the list.

Many global scientific studies reveal that excessive ultraviolet (UV) sunlight and pollution are linked to a decline in amphibian populations. Now Canadian biologists find that too much exposure of excessive UV radiation to tadpole populations reduces their chances of becoming frogs.

2003 marked the 50th anniversary of the research and publication of a different structure of the DNA model proposed by James D. Watson and

Francis H.C. Crick. In 1953 the scientists reported that the DNA molecule resembled a spiral staircase.

A new excavation in South Africa discovered the oldest fossils in the human family. The bones of a skull and a partial arm found in two caves date back to 4 million years ago according to scientists in Johannesburg

Scientists in New Jersey discovered that some outdoor antimosquito coils used to keep insects away can also cause respiratory health problems. The spiral-shaped container releases pollutants in the fumes expelled from coil. The researchers suggest that consumers should check these products carefully.

Researchers in Australia reported that pieces of plastic litter found in oceans continue to have an effect on marine wildlife. Small plastic chips are a hazard for seabirds who mistake the litter for food or fish eggs. The litter also moves up the food chain from fish that have ingested the plastic chips and in turn seals eat them.

2004 A scientific study reported that consumers should limit their consumption of farm-raised Atlantic salmon because of high concentrations of chlorinated organic contaminants in the fish. Their study revealed that the farm-raised salmon were contaminated with polychlorinated biphenyls (PCBs) and other organic chemicals. Except for the PCBs, the researchers agree that the farm-raised fish are healthy but consumption should be limited to no more than once a month in the diet. The researchers based their dietary report on the U.S. Environmental Protection Agency cancer risk assessments.

A group in Salisbury Plain, England is restoring Stonehenge to its natural setting. As a popular historic site to visitors, Stonehenge had become an area surrounded by roads and parking lots. The new restoration plan calls for building an underground tunnel for traffic and removing one of the roads. The present parking lots will become open grassy lawns.

Experts reported that two billion people lack reliable access to safe and nutritious food and 800 million, 40 percent of them children, are classified as chronically malnourished.

Public health officials in Uganda have reported progress in the country's fight against HIV, the AIDS virus. Since 1990's HIV cases in Uganda have dropped by more than 60 percent. Unfortunately, Uganda's neighboring countries are not doing well in their HIV prevention programs.

United Nations Secretary-General Kofi Annan stated, "by 2025, two-thirds of the world's population may be living in countries that face serious water shortages." The growing population is making surface water scarcer particularly in urban areas.

United States and Israel scientists have found a way to produce hydrogen from water. The hydrogen energy can be used in making fuel cells to power vehicles and homes. The research team uses solar radiation to heat sodium hydroxide in a solution of water. At high temperatures the water molecules (H_2O) break apart into oxygen and hydrogen. Using solar-power to produce hydrogen is better environmentally than hydrogen derived from fossil fuels.

Appendix B: Endangered Species by State

The list below, obtained from the U.S. Fish and Wildlife Service, is an abridged listing of a selected group of endangered species (E) for each state. For a full list of endangered and threatened species, and other information about endangered species and the Endangered Species Act, see the Endangered Species Program Website at http://endangered.fws.gov/

ALABAMA

(Alabama has 106 plant and animal species that are listed as endangered (E) or threatened (T). The following list is only a selection of those plants and animals that are endangered. Contact the U.S. Fish and Wildlife Service to see the entire list.)

Animals

E - Bat, gray
E - Bat, Indiana
E - Cavefish, Alabama
T - Chub, spotfin
E - Clubshell, black
E - Combshell, southern
E - Darter, boulder
E - Fanshell
E - Kidneyshell, triangular
E - Lampmussel, Alabama
E - Manatee, West Indian
E - Moccasinshell, Coosa
E - Mouse, Alabama beach
E - Mussel, ring pink
E - Pearlymussel, cracking
E - Pearlymussel, Cumberland monkeyface
E - Pigtoe, dark
E - Plover, piping
E - Shrimp, Alabama cave
E - Snail, tulotoma (Alabama live-bearing)
E - Stork, wood
E - Turtle, Alabama redbelly (red-bellied)
E - Turtle, leatherback sea
E - Woodpecker, red-cockaded

Plants

E - Grass, Tennessee yellow-eyed
E - Leather-flower, Alabama
E - Morefield's leather-flower
E - Pinkroot, gentian
E - Pitcher-plant, Alabama canebrake
E - Pitcher-plant, green
E - Pondberry
E - Prairie-clover, leafy

ALASKA

Animals

E - Curlew, Eskimo (*Numenius borealis*)
E - Falcon, American peregrine (*Falco peregrinus anatum*)

Plant

E - Aleutian shield-fern (Aleutian holly-fern) (*Polystichum aleuticum*)

ARIZONA

Animals

E - Ambersnail, Kanab
E - Bat, lesser (Sanborn's) long-nosed
E - Bobwhite, masked (quail)
E - Chub, bonytail
E - Chub, humpback
E - Chub, Virgin River
E - Chub, Yaqui
E - Flycatcher, Southwestern willow
E - Jaguarundi
E - Ocelot
E - Pronghorn, Sonoran
E - Pupfish, desert
E - Rail, Yuma clapper
E - Squawfish, Colorado
E - Squirrel, Mount Graham red
E - Sucker, razorback
E - Topminnow, Gila (incl. Yaqui)
E - Trout, Gila
E - Vole, Hualapai Mexican
E - Woundfin

Plants

E - Arizona agave
E - Arizona cliffrose
E - Arizona hedgehog cactus
E - Brady pincushion cactus
E - Kearney's blue-star
E - Nichol's Turk's head cactus
E - Peebles Navajo cactus
E - Pima pineapple cactus
E - Sentry milk-vetch

ARKANSAS

Animals

E - Bat, gray
E - Bat, Indiana
E - Bat, Ozark big-eared
E - Beetle, American burying (giant carrion)
E - Crayfish, cave
E - Pearlymussel, Curtis'
E - Pearlymussel, pink mucket
E - Pocketbook, fat
E - Pocketbook, speckled
E - Rock-pocketbook, Ouachita (Wheeler's pearly mussel)
E - Sturgeon, pallid
E - Tern, least
E - Woodpecker, red-cockaded

Plants

E - Harperella
E - Pondberry
E - Running buffalo clover

CALIFORNIA

(California has more than 160 plant and animal species that are listed as endangered or threatened. The following list is only a selection of those plants and animals that are endangered. Contact the U.S. Fish and Wildlife Service to see the entire list.)

Animals

E - Butterfly, El Segundo blue
E - Butterfly, Lange's metalmark
E - Chub, Mohave tui
E - Condor, California
E - Crayfish, Shasta (placid)
E - Fairy shrimp, Conservancy
E - Falcon, American peregrine
E - Fly, Delhi Sands flower-loving
E - Flycatcher, Southwestern willow
E - Fox, San Joaquin kit
E - Goby, tidewater
E - Kangaroo rat, Fresno
E - Lizard, blunt-nosed leopard
E - Mountain beaver, Point Arena
E - Mouse, Pacific pocket
E - Pelican, brown
E - Pupfish, Owens
E - Rail, California clapper
E - Salamander, Santa Cruz long-toed
E - Shrike, San Clemente loggerhead
E - Shrimp, California freshwater
E - Snail, Morro shoulderband (banded dune)
E - Snake, San Francisco garter
E - Stickleback, unarmored threespine
E - Sucker, Lost River
E - Tadpole shrimp, vernal pool
E - Tern, California least
E - Toad, arroyo southwestern
E - Turtle, leatherback sea
E - Vireo, least Bell's
E - Vole, Amargosa

Plants

E - Antioch Dunes evening-primrose
E - Bakersfield cactus
E - Ben Lomond wallflower
E - Burke's goldfields
E - California jewelflower
E - California Orcutt grass
E - Clover lupine
E - Cushenbury buckwheat
E - Fountain thistle
E - Gambel's watercress
E - Kern mallow
E - Loch Lomond coyote-thistle
E - Robust spineflower (includes Scotts Valley spineflower)
E - San Clemente Island larkspur
E - San Diego button-celery
E - San Mateo thornmint
E - Santa Ana River woolly-star
E - Santa Barbara Island liveforever
E - Santa Cruz cypress
E - Solano grass
E - Sonoma sunshine (Baker's stickyseed)

E - Stebbins' morning-glory
E - Truckee barberry
E - Western lily

COLORADO

Animals

E - Butterfly, Uncompahgre fritillary
E - Chub, bonytail
E - Chub, humpback
E - Crane, whooping
E - Ferret, black-footed
E - Flycatcher, Southwestern willow
E - Plover, piping
E - Squawfish, Colorado
E - Sucker, razorback
E - Tern, least
E - Wolf, gray

Plants

E - Clay-loving wild-buckwheat
E - Knowlton cactus
E - Mancos milk-vetch
E - North Park phacelia
E - Osterhout milk-vetch
E - Penland beardtongue

CONNECTICUT

Animals

E - Mussel, dwarf wedge
E - Plover, piping
E - Tern, roseate
E - Turtle, hawksbill sea
E - Turtle, Kemp's (Atlantic) ridley sea
E - Turtle, leatherback sea

Plant

E - Sandplain gerardia

DELAWARE

Animals

E - Plover, piping
E - Squirrel, Delmarva Peninsula fox
E - Turtle, hawksbill sea
E - Turtle, Kemp's (Atlantic) ridley sea—Turtle, green sea
E - Turtle, leatherback sea

Plant

E - Canby's dropwort

FLORIDA

(Florida has more than 90 plant and animal species that are listed as endangered or threatened. The following list is only a selection of those plants and animals that are endangered. Contact the U.S. Fish and Wildlife Service to see the entire list.)

Animals

E - Bat, gray
E - Butterfly, Schaus swallowtail
E - Crocodile, American
E - Darter, Okaloosa
E - Deer, key
E - Kite, Everglade snail
E - Manatee, West Indian (Florida)
E - Mouse, Anastasia Island beach
E - Mouse, Choctawahatchee beach
E - Panther, Florida
E - Plover, piping
E - Rabbit, Lower Keys
E - Rice rat (silver rice rat)
E - Sparrow, Cape Sable seaside
E - Stork, wood
E - Tern, roseate
E - Turtle, hawksbill sea
E - Turtle, Kemp's (Atlantic) ridley sea
E - Turtle, leatherback sea
E - Vole, Florida salt marsh
E - Woodpecker, red-cockaded
E - Woodrat, Key Largo

Plants

E - Apalachicola rosemary
E - Beautiful pawpaw
E - Brooksville (Robins') bellflower
E - Carter's mustard
E - Chapman rhododendron
E - Cooley's water-willow
E - Crenulate lead-plant
E - Etonia rosemary
E - Florida golden aster
E - Fragrant prickly-apple
E - Garrett's mint
E - Key tree-cactus
E - Lakela's mint
E - Okeechobee gourd

E - Scrub blazingstar
E - Small's milkpea
E - Snakeroot
E - Wireweed

GEORGIA

Animals

E - Acornshell, southern
E - Bat, gray
E - Bat, Indiana
E - Clubshell, ovate
E - Clubshell, southern
E - Combshell, upland
E - Darter, amber
E - Darter, Etowah
E - Kidneyshell, triangular
E - Logperch, Conasauga
E - Manatee, West Indian (Florida)
E - Moccasinshell, Coosa
E - Pigtoe, southern
E - Plover, piping
E - Stork, wood
E - Turtle, hawksbill sea
E - Turtle, Kemp's (Atlantic) ridley sea
E - Turtle, leatherback sea
E - Woodpecker, red-cockaded

Plants

E - American chaffseed
E - Black-spored quillwort
E - Canby's dropwort
E - Florida torreya
E - Fringed campion
E - Green pitcher-plant
E - Hairy rattleweed
E - Harperella
E - Large-flowered skullcap
E - Mat-forming quillwort
E - Michaux's sumac
E - Persistent trillium
E - Pondberry
E - Relict trillium
E - Smooth coneflower
E - Tennessee yellow-eyed grass

HAWAII

(Hawaii has 300 plant and animal species listed as endangered or threatened. The following list is only a selection of those plants and animals that are endangered. Contact the U.S. Fish and Wildlife Service to see the entire list.)

Animals

E - 'Akepa, Hawaii (honeycreeper)
E - Bat, Hawaiian hoary
E - Coot, Hawaiian
E - Creeper, Hawaiian
E - Crow, Hawaiian
E - Duck, Hawaiian
E - Duck, Laysan
E - Finch, Laysan (honeycreeper)
E - Finch, Nihoa (honeycreeper)
E - Goose, Hawaiian (nene)
E - Hawk, Hawaiian
E - Millerbird, Nihoa (old world warbler)
E - Nukupu'u (honeycreeper)
E - Palila (honeycreeper)
E - Parrotbill, Maui (honeycreeper)
E - Petrel, Hawaiian dark-rumped
E - Snails, Oahu tree
E - Stilt, Hawaiian
E - Turtle, hawksbill sea
E - Turtle, leatherback sea

Plants

E - Abutilon eremitopetalum
E - Bonamia menziesii
E - Carter's panicgrass
E - Diamond Head schiedea
E - Dwarf iliau
E - Fosberg's love grass
E - Hawaiian bluegrass
E - Hawaiian red-flowered geranium
E - Kaulu
E - Kiponapona
E - Mahoe
E - Mapele
E - Nanu
E - Nehe
E - Opuhe
E - Pamakani
E - Round-leaved chaff-flower
E - Viola helenae

IDAHO

Animals

E - Caribou, woodland
E - Crane, whooping

E - Limpet, Banbury Springs
E - Snail, Snake River physa
E - Snail, Utah valvata
E - Springsnail, Bruneau Hot
E - Springsnail, Idaho
E - Sturgeon, white
E - Wolf, gray

Plants

(No plants on the endangered list)

ILLINOIS

Animals

E - Bat, gray
E - Bat, Indiana
E - Butterfly, Karner blue
E - Dragonfly, Hine's emerald
E - Falcon, American peregrine
E - Fanshell
E - Pearlymussel, Higgins' eye
E - Pearlymussel, orange-foot pimple back
E - Pearlymussel, pink mucket
E, T - Plover, piping
E - Pocketbook, fat
E - Snail, Iowa Pleistocene
E - Sturgeon, pallid
E - Tern, least

Plant

E - Leafy prairie-clover

INDIANA

Animals

E - Bat, gray
E - Bat, Indiana
E - Butterfly, Karner blue
E - Butterfly, Mitchell's satyr
E - Clubshell
E - Fanshell
E - Mussel, ring pink (golf stick pearly)
E - Pearlymussel, cracking
E - Pearlymussel, orange-foot pimple back
E - Pearlymussel, pink mucket
E - Pearlymussel, tubercled-blossom
E - Pearlymussel, white cat's paw
E - Pearlymussel, white wartyback
E - Pigtoe, rough
E, T - Plover, piping
E - Pocketbook, fat
E - Riffleshell, northern
E - Tern, least

Plant

E - Running buffalo clover

IOWA

Animals

E - Bat, Indiana
E - Pearlymussel, Higgins' eye
E - Plover, piping
E - Snail, Iowa Pleistocene
E - Sturgeon, pallid
E - Tern, least

Plants

(No plants on the endangered list)

KANSAS

Animals

E - Bat, gray
E - Bat, Indiana
E - Crane, whooping
E - Curlew, Eskimo
E - Ferret, black-footed
E - Plover, piping
E - Sturgeon, pallid
E - Tern, least
E - Vireo, black-capped

Plants

(No plants on endangered list)

KENTUCKY

Animals

E - Bat, gray
E - Bat, Indiana
E - Bat, Virginia big-eared
E - Clubshell
E - Darter, relict
E - Falcon, American peregrine
E - Fanshell
E - Mussel, ring pink (golf stick pearly)
E - Mussel, winged mapleleaf
E - Pearlymussel, cracking
E - Pearlymussel, Cumberland bean

E - Pearlymussel, dromedary
E - Pearlymussel, little-wing
E - Pearlymussel, orange-foot pimple back
E - Pearlymussel, pink mucket
E - Pearlymussel, purple cat's paw
E - Pearlymussel, tubercled-blossom
E - Pearlymussel, white wartyback
E - Pigtoe, rough
E - Plover, piping
E - Pocketbook, fat
E - Riffleshell, northern
E - Riffleshell, tan
E - Shiner, Palezone
E - Shrimp, Kentucky cave
E - Sturgeon, pallid
E - Tern, least
E - Woodpecker, red-cockaded

Plants

E - Cumberland sandwort
E - Rock cress
E - Running buffalo clover
E - Short's goldenrod

LOUISIANA

Animals

E - Manatee, West Indian (Florida)
E - Pearlymussel, pink mucket
E - Pelican, brown
E - Plover, piping
E - Sturgeon, pallid
E - Tern, least
T - Turtle, green sea
E - Turtle, hawksbill sea
E - Turtle, Kemp's (Atlantic) ridley sea
E - Turtle, leatherback sea
E - Vireo, black-capped
E - Woodpecker, red-cockaded

Plants

E - American chaffseed
E - Louisiana quillwort
E - Pondberry

MAINE

Animals

E - Plover, piping
E - Tern, roseate
E - Turtle, leatherback sea

Plant

E - Furbish lousewort

MARYLAND

Animals

E - Bat, Indiana
E - Darter, Maryland
E - Mussel, dwarf wedge
E - Plover, piping
E - Squirrel, Delmarva Peninsula fox
E - Turtle, hawksbill sea
E - Turtle, Kemp's (Atlantic) ridley sea
E - Turtle, leatherback sea

Plants

E - Canby's dropwort
E - Harperella
E - Northeastern (Barbed bristle) bulrush
E - Sandplain gerardia

MASSACHUSETTS

Animals

E - Beetle, American burying (giant carrion)
E - Falcon, American peregrine
E - Mussel, dwarf wedge
E - Plover, piping
E - Tern, roseate
E - Turtle, hawksbill sea
E - Turtle, Kemp's (Atlantic) ridley sea
E - Turtle, leatherback sea
E - Turtle, Plymouth redbelly (red-bellied)

Plants

E - Northeastern (Barbed bristle)
E - Sandplain gerardia

MICHIGAN

Animals

E - Bat, Indiana
E - Beetle, American burying (giant carrion)
E - Beetle, Hungerford's crawling water
E - Butterfly, Karner blue
E - Butterfly, Mitchell's satyr
E - Clubshell
E - Plover, piping
E - Riffleshell, northern
E - Warbler, Kirtland's
E - Wolf, gray

Plant

E - Michigan monkey-flower

MINNESOTA

Animals

E - Butterfly, Karner blue
E - Mussel, winged mapleleaf
E - Pearlymussel, Higgins' eye
E - Plover, piping
E - Wolf, gray

Plant

E - Minnesota trout lily

MISSISSIPPI

Animals

E - Bat, Indiana
E - Clubshell, black (Curtus' mussel)
E - Clubshell, ovate
E - Clubshell, southern
E - Combshell, southern (penitent mussel)
E - Crane, Mississippi sandhill
E - Falcon, American peregrine
E - Manatee, West Indian (Florida)
E - Pelican, brown
E - Pigtoe, flat (Marshall's mussel)
E - Pigtoe, heavy (Judge Tait's mussel)
E - Plover, piping
E - Pocketbook, fat
E - Stirrupshell
E - Sturgeon, pallid
E - Tern, least
E - Turtle, hawksbill sea
E - Turtle, Kemp's (Atlantic) ridley sea
E - Turtle, leatherback sea
E - Woodpecker, red-cockaded

Plants

E - American chaffseed
E - Pondberry

MISSOURI

Animals

E - Bat, gray
E - Bat, Indiana
E - Bat, Ozark big-eared
E - Pearlymussel, Curtis'
E - Pearlymussel, Higgins' eye
E - Pearlymussel, pink mucket
E - Plover, piping
E - Pocketbook, fat
E - Sturgeon, pallid
E - Tern, least

Plants

E - Missouri bladderpod
E - Pondberry
E - Running buffalo clover

MONTANA

Animals

E - Crane, whooping
E - Curlew, Eskimo
E - Ferret, black-footed
E - Plover, piping
E - Sturgeon, pallid
E - Sturgeon, white
E - Tern, least
E - Wolf, gray

Plants

(No plants on endangered list)

NEBRASKA

Animals

E - Beetle, American burying
(giant carrion)
E - Crane, whooping
E - Curlew, Eskimo
E - Ferret, black-footed
E - Plover, piping
E - Sturgeon, pallid
E - Tern, least

Plant

E - Blowout penstemon

NEVADA

Animals

E - Chub, bonytail
E - Chub, Pahranagat roundtail (bonytail)
E - Chub, Virgin River
E - Cui-ui
E - Dace, Ash Meadows speckled

E - Dace, Clover Valley speckled
E - Dace, Independence Valley speckled
E - Dace, Moapa
E - Poolfish (killifish), Pahrump
E - Pupfish, Ash Meadows Amargosa
E - Pupfish, Devils Hole
E - Pupfish, Warm Springs
E - Spinedace, White River
E - Springfish, Hiko White River
E - Springfish, White River
E - Sucker, razorback
E - Woundfin

Plants

E - Amargosa niterwort
E - Steamboat buckwheat

NEW HAMPSHIRE

Animals

E - Butterfly, Karner blue
E - Mussel, dwarf wedge
E - Turtle, leatherback sea

Plants

E - Jesup's milk-vetch
E - Northeastern (Barbed bristle) bulrush
E - Robbins' cinquefoil

NEW JERSEY

Animals

E - Bat, Indiana
E - Plover, piping
E - Tern, roseate
E - Turtle, hawksbill sea
E - Turtle, Kemp's (Atlantic) ridley sea
E - Turtle, leatherback sea

Plant

E - American chaffseed

NEW MEXICO

Animals

E - Bat, lesser (Sanborn's) long-nosed
E - Bat, Mexican long-nosed
E - Crane, whooping
E - Gambusia, Pecos
E - Isopod, Socorro
E - Minnow, Rio Grande silvery
E - Springsnail, Alamosa
E - Springsnail, Socorro
E - Sucker, razorback
E - Tern, least
E - Topminnow, Gila (incl. Yaqui)
E - Trout, Gila
E - Woundfin

Plants

E - Holy Ghost ipomopsis
E - Knowlton cactus
E - Kuenzler hedgehog cactus
E - Lloyd's hedgehog cactus
E - Mancos milk-vetch
E - Sacramento prickly-poppy
E - Sneed pincushion cactus
E - Todsen's pennyroyal

NEW YORK

Animals

E - Butterfly, Karner blue
E - Mussel, dwarf wedge
E, T - Plover, piping
E - Tern, roseate
E - Turtle, hawksbill sea
E - Turtle, Kemp's (Atlantic) ridley sea
E - Turtle, leatherback sea

Plants

E - Northeastern (Barbed bristle) bulrush
E - Sandplain gerardia

NORTH CAROLINA

Animals

E - Bat, Indiana
E - Bat, Virginia big-eared
E - Butterfly, Saint Francis' satyr
E - Elktoe, Appalachian
E - Falcon, American peregrine
E - Heelsplitter, Carolina
E - Manatee, West Indian (Florida)
E - Mussel, dwarf wedge
E - Pearlymussel, little-wing
E - Plover, piping
E - Shiner, Cape Fear
E - Spider, spruce-fir moss

E - Spinymussel, Tar River
E - Squirrel, Carolina northern flying
E - Tern, roseate
E - Turtle, hawksbill sea
E - Turtle, Kemp's (Atlantic) ridley sea
E - Turtle, leatherback sea
E - Wolf, red
E - Woodpecker, red-cockaded

Plants

E - American chaffseed
E - Bunched arrowhead
E - Canby's dropwort
E - Cooley's meadowrue
E - Green pitcher-plant
E - Harperella
E - Michaux's sumac
E - Mountain sweet pitcher-plant
E - Pondberry
E - Roan Mountain bluet
E - Rock gnome lichen
E - Rough-leaved loosestrife
E - Schweinitz's sunflower
E - Small-anthered bittercress
E - Smooth coneflower
E - Spreading avens
E - White irisette

NORTH DAKOTA

Animals

E - Crane, whooping
E - Curlew, Eskimo
E - Falcon, American peregrine
E - Ferret, black-footed
E - Plover, piping
E - Sturgeon, pallid
E - Tern, least
E - Wolf, gray

Plants

(No plants on endangered list)

OHIO

Animals

E - Bat, Indiana
E - Beetle, American burying (giant carrion)
E - Butterfly, Karner blue
E - Butterfly, Mitchell's satyr
E - Clubshell
E - Dragonfly, Hine's emerald
E - Fanshell
E - Madtom, Scioto
E - Pearlymussel, pink mucket
E - Pearlymussel, purple cat's paw
E - Pearlymussel, white cat's paw
E, T - Plover, piping
E - Riffleshell, northern

Plant

E - Running buffalo clover

OKLAHOMA

Animals

E - Bat, gray
E - Bat, Indiana
E - Bat, Ozark big-eared
E - Beetle, American burying (giant carrion)
E - Crane, whooping
E - Curlew, Eskimo
E - Plover, piping
E - Rock-pocketbook, Ouachita
E - Tern, least
E - Vireo, black-capped
E - Woodpecker, red-cockaded

Plants

(No plants on the endangered list)

OREGON

Animals

E - Chub, Borax Lake
E - Chub, Oregon
E - Deer, Columbian white-tailed
E - Pelican, brown
E - Sucker, Lost River
E - Sucker, shortnose
E - Turtle, leatherback sea

Plants

E - Applegate's milk-vetch
E - Bradshaw's desert-parsley
E - Malheur wire-lettuce
E - Marsh sandwort
E - Western lily

PENNSYLVANIA

Animals

E - Bat, Indiana
E - Clubshell
E - Mussel, dwarf wedge
E - Mussel, ring pink (golf stick pearly)
E - Pearlymussel, cracking
E - Pearlymussel, orange-foot pimple back
E - Pearlymussel, pink mucket
E - Pigtoe, rough
E,T - Plover, piping
E - Riffleshell, northern

Plant

E - Northeastern (Barbed bristle) bulrush

RHODE ISLAND

Animals

E - Beetle, American burying
E - Falcon, American peregrine
E - Plover, piping
E - Tern, roseate
E - Turtle, hawksbill sea
E - Turtle, Kemp's
E - Turtle, leatherback sea

Plant

E - Sandplain gerardia

SOUTH CAROLINA

Animals

E - Bat, Indiana
E - Heelsplitter, Carolina
E - Manatee, West Indian (Florida)
E - Plover, piping
E - Stork, wood
E - Tern, roseate
E - Turtle, hawksbill sea
E - Turtle, Kemp's (Atlantic) ridley sea
E - Turtle, leatherback sea
E - Woodpecker, red-cockaded

Plants

E - American chaffseed
E - Black-spored quillwort
E - Bunched arrowhead
E - Canby's dropwort
T - Dwarf-flowered heartleaf
E - Harperella
E - Michaux's sumac
E - Mountain sweet pitcher-plant
E - Persistent trillium
E - Pondberry
E - Relict trillium
E - Rough-leaved loosestrife
E - Schweinitz's sunflower
E - Smooth coneflower

SOUTH DAKOTA

Animals

E - Beetle, American burying (giant carrion)
E - Crane, whooping
E - Curlew, Eskimo
E - Ferret, black-footed
E - Plover, piping
E - Sturgeon, pallid
E - Tern, least
E - Wolf, gray

Plants

(No plants on the endangered list)

TENNESSEE

(Tennessee has 81 plant and animal species that are listed as endangered or threatened. The following list is only a selection of those plants and animals that are endangered. Contact the U.S. Fish and Wildlife Service to see the entire list.)

Animals

E - Bat, gray
E - Bat, Indiana
E - Combshell, upland
E - Crayfish, Nashville
E - Darter, amber
E - Fanshell
E - Lampmussel, Alabama
E - Madtom, Smoky
E - Marstonia (snail), (royalobese)
E - Moccasinshell, Coosa
E - Mussel, ring pink (golf stick pearly)
E - Pearlymussel, Appalachian monkeyface

E - Pearlymussel, Cumberland bean
E - Riversnail, Anthony's
E - Spider, spruce-fir moss
E - Squirrel, Carolina northern flying
E - Sturgeon, pallid
E - Tern, least
E - Wolf, red
E - Woodpecker, red-cockaded

Plants

E - Cumberland sandwort
E - Green pitcher-plant
E - Large-flowered skullcap
E - Leafy prairie-clover (Dalea)
E - Roan Mountain bluet
E - Rock cress
E - Rock gnome lichen
E - Ruth's golden aster
E - Spring Creek bladderpod
E - Tennessee purple coneflower
E - Tennessee yellow-eyed grass

TEXAS

(Texas has 70 plant and animal species that are listed as endangered or threatened. The following list is only a selection of those plants and animals that are endangered. Contact the U.S. Fish and Wildlife Service to see the entire list.)

Animals

E - Bat, Mexican long-nosed
E - Beetle, Coffin Cave mold
E - Crane, whooping
E - Curlew, Eskimo
E - Darter, fountain
E - Falcon, northern aplomado
E - Jaguarundi
E - Manatee, West Indian (Florida)
E - Minnow, Rio Grande silvery
E - Ocelot
E - Pelican, brown
E - Plover, piping
E - Prairie-chicken, Attwater's greater
E - Pupfish, Comanche Springs
E - Salamander, Texas blind
E - Spider, Tooth Cave
E - Tern, least
E - Toad, Houston
E - Turtle, hawksbill sea
E - Turtle, Kemp's (Atlantic) ridley sea
E - Vireo, black-capped
E - Warbler, golden-cheeked
E - Woodpecker, red-cockaded

Plants

E - Ashy dogweed
E - Black lace cactus
T - Hinckley's oak
E - Large-fruited sand-verbena
E - Little Aguja pondweed
E - Lloyd's hedgehog cactus
E - Nellie cory cactus
E - Sneed pincushion cactus
E - South Texas ambrosia
E - Star cactus
E - Terlingua Creek cats-eye
E - Texas poppy-mallow
E - Texas snowbells
E - Texas wild-rice
E - Tobusch fishhook cactus
E - Walker's manioc

UTAH

Animals

E - Ambersnail, Kanab
E - Chub, bonytail
E - Chub, humpback
E - Chub, Virgin River
E - Crane, whooping
E - Ferret, black-footed
E - Flycatcher, Southwestern willow
E - Snail, Utah valvata
E - Squawfish, Colorado
E - Sucker, June
E - Sucker, razorback
E - Woundfin

Plants

E - Autumn buttercup
E - Barneby reed-mustard
E - Barneby ridge-cress (peppercress)
E - Clay phacelia
E - Dwarf bear-poppy
E - Kodachrome bladderpod
E - San Rafael cactus
E - Shrubby reed-mustard (toad-flax cress)
E - Wright fishhook cactus

VERMONT

Animals

E - Bat, Indiana
E - Mussel, dwarf wedge

Plants

E - Jesup's milk-vetch
E - Northeastern (Barbed bristle) bulrush

VIRGINIA

Animals

E - Bat, gray
E - Bat, Indiana
E - Bat, Virginia big-eared
E - Darter, duskytail
E - Falcon, American peregrine
E - Fanshell
E - Isopod, Lee County cave
E - Logperch, Roanoke
E - Mussel, dwarf wedge
E - Pearlymussel, Appalachian monkeyface
E - Pearlymussel, birdwing
E - Pearlymussel, cracking
E - Pearlymussel, Cumberland monkeyface
E - Pearlymussel, dromedary
E - Pearlymussel, green-blossom
E - Pearlymussel, little-wing
E - Pearlymussel, pink mucket
E - Pigtoe, fine-rayed
E - Pigtoe, rough
E - Pigtoe, shiny
E - Plover, piping
E - Riffleshell, tan
E - Salamander, Shenandoah
E - Snail, Virginia fringed mountain
E - Spinymussel, James River (Virginia)
E - Squirrel, Delmarva Peninsula fox
E - Squirrel, Virginia northern flying
E - Turtle, hawksbill sea
E - Turtle, Kemp's (Atlantic) ridley sea
E - Turtle, leatherback sea
E - Woodpecker, red-cockaded

Plants

E - Northeastern (Barbed bristle) bulrush
E - Peter's Mountain mallow
E - Shale barren rock-cress
E - Smooth coneflower

WASHINGTON

Animals

E - Caribou, woodland
E - Deer, Columbian white-tailed
E - Pelican, brown
E - Turtle, leatherback sea
E - Wolf, gray

Plants

E - Bradshaw's desert-parsley (lomatium)
E - Marsh sandwort

WEST VIRGINIA

Animals

E - Bat, Indiana
E - Bat, Virginia big-eared
E - Clubshell
E - Falcon, American peregrine
E - Fanshell
E - Mussel, ring pink
E - Pearlymussel, pink mucket
E - Pearlymussel, tubercled-blossom
E - Riffleshell, northern
E - Spinymussel, James River
E - Squirrel, Virginia northern flying

Plants

E - Harperella
E - Northeastern (Barbed bristle) bulrush
E - Running buffalo clover
E - Shale barren rock-cress

WISCONSIN

Animals

E - Butterfly, Karner blue
E - Dragonfly, Hine's emerald
E - Mussel, winged mapleleaf
E - Pearlymussel, Higgins' eye
E, T - Plover, piping
E - Warbler, Kirtland's
E - Wolf, gray

Plants

(No plants on endangered list)

WYOMING

Animals

E - Crane, whooping
E - Dace, Kendall Warm Springs
E - Ferret, black-footed
E - Squawfish, Colorado
E - Sucker, razorback
E - Toad, Wyoming
E - Wolf, gray

Plants

(No plants on endangered list)

APPENDIX C: WEBSITES BY CLASSIFICATION

Please note that the authors have made a consistent effort to include up-to-date Websites. However, over time, some Websites may move or no longer be posted.

ACID MINE DRAINAGE

National Reclamation Center, West Virginia University, Evansdale office, http//www.nrcce.wvu.edu/

ACID RAIN

http://www.epa.gov/docs/acidrain/andhome/html.

The EPA has a hotline to request educational materials or respond to questions regarding acid rain: (202) 343–9620. http://www.econet.apc.org/acid rain.

Environmental Protection Agency, http://www.epa.gov/docs/acidrain/effects/enveffct.html.

National Reclamation Center's West Virginia University, Evansdale office: http://www.nrcce.wvu.edu/

USGS Water Science/Acid Rain, http://wwwga.usgs.gov/edu/acidrain.html.

AGENCY FOR TOXIC SUBSTANCES AND DISEASES

Registry Division of Toxicology
1600 Clifton Road NE Mailstop E-29
Atlanta, GA 30333
Website: http://www.atsdr1.atsdr.cdc.gov:8080/atsdrhome.html.

Agency for Toxic Substances and Disease Registry, http://www.atsdr.cdc.gov/cxcx3.html.

Information on biosphere reserves and UNESCO's Man and the Biosphere Programme, UNESCO: http://www.unesco. org

Man and the Biosphere Program: http://www. mabnet.org

AGRICULTURE

United States Department of Agriculture, http://www.usda.gov.

ALTERNATIVE FUELS

Department of Energy, http://www.doe.gov.

Department of Energy Alternative Fuels Data Center, http://www.afdc.nrel.gov; http://www.afdc.doe.gov/; or http://www.fleets.doe.gov.

AMPHIBIANS

http://www.frogweb.gov/

ANTARCTICA

Antarctica Treaty, http://www.sedac.ciesin.org/pidb/register/reg-024.rrr.html.

Greenpeace International Antarctic Homepage, http://www.greenpeace.org/~comms/98/antarctic.

International Centre for Antarctic Information and Research Homepage (includes text of Antarctic Treaty), http://www.icair.iac.org.nz.

Virtual Antarctica, http://www.exploratorium.edu

ARCTIC

Arctic Circle (University of Connecticut), http://arcticcircle.uconn.edu/arcticcircle.

Arctic Council Home Page, http://www.nrc.ca/arctic/index.html.

Arctic Monitoring and Assessment Programme (Norway), http://www.gsf.de/ UNEP/amap1.html.

Arctic National Wildlife Refuge, http://energy.usgs.gov/factsheets/ANWR/ANWR.html.

Institute of Arctic and Alpine Research, http://instaar.colorado.edu.

Institute of the North (Alaska Pacific University),

Inuit Circumpolar Conference,
NOAA Fisheries, http://www.nmfs.gov/.

Nunavut,

Smithsonian Institution Arctic Studies Center, http://www.mnh.si.edu/arctic.

U.S. Fish and Wildlife Service
U.S. Department of the Interior

1849 C Street, NW,
Washington, D.C. 20240
Telephone: (202) 208-5634
Website: http://www.fws.gov.

World Conservation Monitoring Centre Arctic Programme, http://www.wcmc.org.uk/arctic.

AUTOMOBILE

Cars and Their Enviromental Impact, http://www.environment.volvocars.com/ch1-1.htm.

National Center for Vehicle Emissions Control and Safety (NCVECS), http://www.colostate.edu/Depts/NCVECS/ncvecs1.html.

U.S. Environmental Protection Agency Fact Sheet (EPA 400-F-92-004, August 1994), "Air Toxics from Motor Vehicles," http://www.epa.gov/oms/02-toxic.htm.

U.S. Enviromental Protection Agency, Office of Mobile Sources, http://www.epa.gov/oms.

BIOLOGICAL WEAPONS

Federation of American Scientist Biological Weapons Control, http://www.fas.org/bwc.

Chemical and Biological Defense Information Analysis Center, http://www.cbiac.apgea.army. mil

BIOMES

Committee for the National Institute for the Environment, http://www.cnie.org/nle/biodv-6.html.

BIOREMEDIATION

Consortium, http://www.rtdf.org/public/biorem.

BROWNFIELD

Projects, http://www.epa.gov/brownfields/.

CERES

Website: http://www.ceres.org or e-mail ceres@igc.apc.org.

Summaries of Major Environmental Laws, http://www.epa.gov/region5/defs/index.html.

CHEETAHS

Cheetah Conservation Fund

4649 Sunnyside Avenue N, Suite 325
Seattle, WA 98103
Website: http://www.cheetah.org.

World Wildlife Fund

1250 24th Street, NW,
Washington, D.C. 20037
Telephone: 1-800-225-5993
Website: http://www.worldwildlife.org/.

CHEMICAL WEAPONS

Chemical Stockpile Disposal Project (CSDP), http://www.pmcd.apgea.army.mil/graphical/CSDP/index.html.

Tooele Chemical Agent Disposal Site Facility, http://www.deq.state.ut.us/eqshw/cds/tocdfhp1.htm.

CLEAN WATER ACT

Sierra Club, "Happy 25th Birthday, Clean Water Act," http://sierraclub.org/wetlands/cwabday.html.

CLIMATE CHANGE AND GLOBAL WARMING

U.S. Geological Survey, Climate Change and History, http://geology.usgs.gov/index.shtml.

EPA Global Warming Site, http://www.epa.gov/globalwarming.

Greenpeace International, Climate, http://www.greenpeace.org/~climate.

United Nations Intergovernmental Panel on Climate Change, http://www.ipcc.ch.

COAL

Coal Age Magazine, http://coalage.com.

Department of Energy, Office of Fossil Energy, http:/www.doe.gov.

U.S. Geological Survey, National Coal Resources Data System, http:energy.er.usgs.gov/coalqual. htm.

COASTAL AND MARINE GEOLOGY

U.S. Geological Survey, http://marine.usgs.gov/.

COMPOSTING

EPA Office of Solid Waste and Emergency Response—Composting, http:www.epa.gov/epaoswer/non-hw/compost/index.htm

Cornell Composting, http://www.cfe.cornell.edu/compost/Composting_Homepage.html

CONSENT DECREES

EPA Office of Enforcement and Compliance Assurance, http://es.epa.gov/oeca/osre/decree.html.

CORAL REEFS

Coral Reef Alliance, http://www.coral.org.

Coral Reef Network Directory, Greenpeace
1436 U Street, NW
Washington, D.C. 20009
Website: http://www.greenpeace.org.

EARTHDAY 2000

Earth Day Network

91 Marion Street,
Seattle, WA 98104
Telephone: 1(206)-264-0114.
Website: http://www.earthday.net/; and worldwide@earthday.net.

EARTHWATCH

Earthwatch Institute International, http://www. earthwatch.org.

EL NIÑO

El Niño/La Niña theme page, contact NOAA Website: http://www.pmel.noaa.gov/toga-tao/el-nino/nino-home-low.html.

NOAA, La Niña homepage, www.elnino.noaa.gov/lanina.html.

National Center for Atmospheric Research, http://www.ncar.ucar.edu/.

National Hurricane Center/Tropical Prediction Center, http://www.nhc.noaa.gov/.

National Oceanographic and Atmospheric Administration, http://www.noaa.gov/.

Scripps Institute of Oceanography, http://sio.ucsd.edu/supp_groups/siocomm/elnino/elnino.html.

ELECTRIC VEHICLES

Electric Vehicle Association of the Americas 800-438-3228, http://www.evaa.org.

Electric Vehicle Technology, http://www.avere.org/.

ELEPHANTS

African Wildlife Foundation, http://www.awf.org.

U.S. Fish and Wildlife Service, Species List of Endangered and Threatened Wildlife, http://endangered.fws.gov/

World Wildlife Fund, http://www.wwf.org.

ETHANOL

U.S. Department of Energy, Energy Efficiency and Renewable Energy Clearinghouse,

P.O. Box 3048
Merrifield, VA 22116
E-mail: energyinfo@delphi.com.
Website: http://www.doe.gov.

EVERGLADES

National Park Service, Everglades National Park, http://www.nps.gov/ever.

FEDERAL EMERGENCY MANAGEMENT AGENCY (FEMA)

FEMA, http://www.fema.gov.

FISHING, COMMERCIAL

National Oceanographic and Atmospheric Administration Fisheries, http://www.nmfs. gov/.

United Nations Food and Agriculture Organization Fisheries, http://www.fao.org/waicent/faoinfo/fishery/fishery.htm.

FORESTS

American Forests, http://www.amfor.org.

Greenpeace International, Forests, http://www. greenpeace.org/~forests.

Society of American Foresters, http://www. safnet.org.

U.S. Forest Service, http://www.fs.fed.us.

U.S. Forest Service Research, http://www.fs.fed.us/links/research.shtml.

World Conservation Monitoring Centre, http://www.wcmc.org.uk.

World Resources Institute Forest Frontiers Initiative, http://www.wri.org/ffi.

World Wildlife Fund (Worldwide Fund for Nature) Forests for Life Campaign, http://www.panda.org/forests4life.

FUEL CELLS AND OTHER ALTERNATIVE FUELS

Crest's Guide to the Internet's Alternative Energy Resources, http://solstice.crest.org/online/aeguide/aehome.html.

U.S. Department of Energy

P.O. Box 12316
Arlington, VA 22209
Telephone: 1-800-423-1363
Website: http://www.doe.gov.

U.S. Department of Energy, Alternative Fuels Data Center, http://www.afdc.nrel.gov.

GEOLOGY

Geological surveys, U.S. Geological Survey, http://www.usgs.gov/.

For general interest publications and products, http://mapping.usgs.gov/www/products/mappubs.html.

GEOTHERMAL SITES

Energy and Geoscience Institute

University of Utah
423 Wakara Way
Salt Lake City, UT 84108
Website: http://www.egi.utah.edu.

Geothermal energy information, http://geothermal.marin.org.

Geothermal database USA and Worldwide, http://www.geothermal.org.

International geothermal, http://www.demon.co.uk/geosci/igahome.html.

Solstice is the Internet information service of the Center for Renewable Energy and Sustainable Technology (CREST), http://solstice.crest.org/

GLACIERS SHRINKING

United States Geological Survey, Climate Change and History, http://geology.usgs.gov/index. shtml.

Sierra Club, Public Information Center, (415) 923-5653; or the Global Warming and Energy Team, (202) 547-1141, or by E-mail: information@sierraclub.org.

GLOBEC

Educational Website, http://cbl.umces.edu/fogarty/usglobec/misc/education.html.

GRASSLANDS AND PRAIRIES

Postcards from the Prairie, http://www.nrwrc.usgs.gov/postcards/postcards.htm.

University of California, Berkeley, World Biomes, Grasslands, http://www.ucmp.berkeley.edu/glossary/gloss5/biome/grasslan.html.

Worldwide Fund for Nature, Grasslands and Its Animals, http://www.panda.org/kids/wildlife/idxgrsmn.htm.

GROUNDWATER

EPA, http://www.epa.gov/swerosps/ej/.

Groundwater atlas of the United States, http://www.capp.er.usgs.gov/publicdocs/gwa/.

HAZARDOUS MATERIALS TRANSPORTAION ACT

Website: http://www.dot.gov.

HAZARDOUS SUBSTANCES

U.S. Environmental Protection Agency Program, http://epa.gov/.

U.S. Occupational Safety and Health Administration (OSHA), http://www.osha.gov/toxicsubstances/index.html.

Environmental Defense Fund (data on wastes and chemicals at U.S. sources), http://www.scorecard.org.

HAZARDOUS WASTE TREATMENT

Federal Remedial Technologies Roundtable, Hazardous Waste Clean-Up Information ("CLU-IN"), http://www.clu-in.org.

HEAVY METALS

U.S. Environmental Protection Agency, Office of Pollution Prevention and Toxics, http://www.epa.gov/opptintr.

HIGH-LEVEL RADIOACTIVE WASTES

U.S. Nuclear Regulatory Commission, Radioactive Waste Page, http://www.nrc.gov/NRC/radwaste.

U.S. Environmental Protection Agency, Mixed-Waste Homepage, http://www.epa.gov/radiation/mixed-waste.

HURRICANES

National Hurricane Center, http://www.nhc.noaa.gov.

HYDROELECTRIC POWER

U.S. Bureau of Reclamation Hydropower Information, http://www.usbr.gov/power/edu/edu.htm.

U.S. Geological Survey, http://wwwga.usgs.gov/edu/hybiggest.html.

HYDROGEN

National Renewable Energy Laboratory, http://www.nrel.gov/lab/pao/hydrogen.html.

EnviroSource, Hydrogen InfoNet, http:///www.eren.doe.gov/hydrogen/infonet.html.

INTERNATIONAL ATOMIC ENERGY AGENCY

Agency, http://www.iaea.org.

Managing Radioactive Waste Fact Sheet, http://www.iaea.org/worldatom/inforesource/factsheets/manradwa.html.

INTERNATIONAL COUNCIL FOR LOCAL ENVIRONMENTAL INITIATIVES

Homepage, http://www.iclei.org.

INTERNATIONAL REGISTER OF POTENTIALLY TOXIC CHEMICALS

Homepage, http://www.unep.org/unep/program/hhwb/chemical/irptc/home.htm.

INTERNATIONAL WHALING COMMISSION

Homepage, http://www.ourworld.compuserve.com/homepages/iwcoffice.

INVERTEBRATES: THREATENED AND ENDANGERED

U.S. Fish and Wildlife Service, Species List of Endangered and Threatened Wildlife, http://endangered.fws.gov/

LANDSAT AND SATELLITE IMAGES

Earthshots, Satellite Images of Environmental Change, http://www.usgs.gov/Earthshots/.

Landsat Gateway, http://landsat.gsfc.nasa.gov/main.htm.

LEAD

National Lead Information Center's Clearinghouse, 1-800-424-LEAD, http://www.epa.gov/lead/.

LEOPARDS

U.S. Fish and Wildlife Service, Species List of Endangered and Threatened Wildlife, http://www.fws.gov/r9endspp/lsppinfo.html.

LITTER

Keep America Beautiful, http://www.kab.org.

MAMMALS

U.S. Fish and Wildlife Service, Vertebrate Animals, http://www.fws.gov/r9endspp/lsppinfo.html.

MANATEES

Save the Manatees, http://www.savethemanatee. org.

Sea World, Manatees, http://www.seaworld.org/manatee/sciclassman.html.

MARSHES

Environmental Protection Agency, Office of Wetlands, Oceans, Watersheds, http://www.epa.gov/owow/wetlands/wetland2.html.

North American Waterfowl and Wetlands Office, http://www.fws.gov/r9nawwo.

North American Wetlands Conservation Act, http://www.fws.gov/r9nawwo/nawcahp.html.

North American Wetlands Conservation Council, http://www.fws.gov/r9nawwo/nawcc.html.

Wetlands, wetlands-hotline@epamail.epa.gov.

MATERIAL SAFETY DATA SHEET

Toxic chemicals, http://www.siri.org/msds; http://www.ilpi.com/mads/index.html.

MENDES, CHICO

Chico Mendes, http://www.edf.org/chico.

NATURAL DISASTERS

Building Safer Structures, http://quake.wr.usgs. gov/QUAKES/FactSheets/SaferStructures/.

Center for Integration of Natural Disaster Information, http://cindi.usgs.gov/events/.

Earthquakes, http://quake.wr.usgs.gov/; http://geology.usgs.gov/quake.html. For the latest earthquake information http://quake.wr.usgs.gov/QUAKES/CURRENT/current.html

National Hurricane Center, http://www.nhc.noaa.gov.

U.S. Geological Survey, http://geology.usgs.gov/whatsnew.html.

NATIONAL MARINE FISHERIES

History of National Marine Fisheries Service, http://www.wh.whoi.edu/125th/history/century.html.

National Marine Fisheries, http://kingfish.ssp.nmfs.gov.

NOAA Fisheries, http://www.nmfs.gov/.

NATIONAL OCEAN AND ATMOSPHERIC ADMINISTRATION (NOAA)

Climate forecasting, http://www.cdc.noaa.gov/ Seasonal/.

El Niño Theme Page, http://www.pmel.noaa.gov/toga-tao/el-nino/nino-home-low.html.

Homepage, http://www.noaa.gov/.

Recover Protected Species, http://www.noaa.gov/nmfs/recover.html.

Safe Navigation Page, http://anchor.ncd.noaa.gov/psn/psn.htm.

NATIONAL WEATHER SERVICE

Homepage, http://www.nws.noaa.gov.

NATIONAL WILDLIFE REFUGE SYSTEM

Homepage, http://refuges.fws.gov/NWRSHomePage.html.

NATURAL GAS

American Gas Association, http://www.aga.org.

Oil and Gas Journal Online, http://www.ogjonline.com.

U.S. Department of Energy, Energy Information Administration, http://www.eia.doe.gov.

U.S. Department of Energy, Office of Fossil Energy, http://www.fe.doe.gov.

U.S. Geological Survey Energy, Resources Program, http://energy.usgs.gov/index.html.

NOISE POLLUTION

Noise Pollution Clearinghouse, http://www. nonoise.org.

NONPOINT SOURCES

Nonpoint Source Pollution Control Program, http://www.epa.gov/OWOW/NPS/whatudo.html; http://www.epa.gov/OWOW/ NPS/.

NUCLEAR ENERGY AND NUCLEAR REACTORS

American Nuclear Society, http://www.ans.org.

Nuclear Energy Institute, http://www.nei.org.

Nuclear Information and Resource Service, http://www.nirs.org.

U.S. Department of Energy, Office of Nuclear Energy, Science and Technology, http://www.ne.doe.gov.

U.S. Nuclear Regulatory Commission, http://www.nrc.gov.

NUCLEAR WASTE POLICY ACT

American Nuclear Society, http://www.ans.org.

Nuclear Energy Institute, http://www.nei.org.

NUCLEAR WASTE SITES

Hazard Ranking System, http://www.epa.gov/ superfund/programs/npl_hrs/hrsint.htm.

National Research Council, Board on Radioactive Waste Management, http://www4.nas.edu/brwm/brwm-res.nsf.

Superfund, http://www.pin.org/superguide.htm; http://www.epa.gov/superfund.

U.S. Department of Energy, Office of Civilian Radioactive Waste Management, http://www.rw.doe.gov.

U.S. Environmental Protection Agency, Mixed-Waste Homepage, http://www.epa.gov/radiation/mixed-waste.

U.S. Nuclear Regulatory Commission, Radioactive Waste Page, http://www.nrc.gov/ NRC/radwaste.

OCCUPATIONAL SAFETY AND HEALTH ACT (OSHA)

OSHA Homepage, http://www.osha.gov.

OCEAN THERMAL ENERGY CONVERSION (OTEC)

National Renewable Energy Laboratory

1617 Cole Boulevard
Golden, CO 80401
Website: http:llnrelinfo.nrel.gov.

Natural Energy Laboratory of Hawaii, http://bigisland.com/nelha/index.html.

OCEANS

National Oceanographic and Atmospheric Administration, http://www.noaa.gov/.

Safe Ocean Navigation Page, http://anchor.ncd.noaa.gov/psn/psn.htm.

OFFICE OF SURFACE MINING

Office of Surface Mining, http://www.osmre.gov.

Appalachian Clean Streams Initiative, majordomo@osmre.gov.

OLD-GROWTH FORESTS

Greenpeace International, Forests, http://www.greenpeace.org/~forests.

World Resources Institute, Forest Frontiers Initiative, http://www.wri.org/ffi.

OLMSTEAD, FREDERICK LAW

Homepage, http://fredericklawolmsted.com.

ORGANIZATION OF PETROLEUM EXPORTING COUNTRIES (OPRC)

Homepage, http://www.opec.org.

OVERFISHING

Information and data statistics, http://www.nmfs. gov.

National Aeronautics and Space Administration, Ocean Planet, http://seawifs.gsfc.nasa.gov/OCEAN_PLANET/HTML/peril_overfishing.html.

National Marine Fisheries Service, http://www. nmfs.gov.

NOAA, http://www.noaa.gov.

United Nations Food and Agricultural Organization, http://www.fao.org.

United Nations Food and Agriculture Organization Fisheries, http://www.fao.org/.

United Nations System, http://www.unsystem.org.

OZONE-RELATED ISSUES

Environmental Protection Agency, science of ozone depletion, http://www.epa.gov/ozone/science/.

NOAA, Commonly Asked Questions about Ozone, www.publicaffairs.noaa.gov/grounders/ozo1.html.

NOAA, Network for the Detection of Stratospheric Change, www.noaa.gov.

PARROTS

Online Book of Parrots, http://www.ub.tu-clausthal.dep/p_welcome.html.

World Parrot Trust, http://www.worldparrottrust.org.

World Wildlife Fund, http:www.panda.org.

PESTICIDES

Toxics and Pesticides, http://www.epa.gov/oppfead1/work_saf/.

Pesticides in the Atmosphere, http://ca.water.usgs.gov/pnsp/atmos.

PETERSON, ROGER TORY

Roger Tory Peterson Institute of Natural History,

311 Curtis Street
Jamestown, NY 14701
Website: http://www.rtpi.org/info/rtp.htm.

PETROLEUM

American Petroleum Institute, http://www.api.org.

Petroleum Information, http://www.petroleuminformation.com.

Oil and Gas Journal Online, http://www. ogjonline.com.

U.S. Department of Energy, Energy Information Administration, http://www.eia.doe.gov.

U.S. Department of Energy, Office of Fossil Energy, http://www.fe.doe.gov.

U.S. Geological Survey Energy Resources Program, http://energy.usgs.gov/index.html.

U.S. Geological Survey Fact Sheet FS-145-97, "Changing Perceptions of World Oil and Gas Resources as Shown by Recent USGS Petroleum Assessments," http://greenwood.cr.usgs.gov/pub/fact-sheets/fs-0145-97/fs-0145-97.html.

PLUTONIUM

U.S. Nuclear Regulatory Commission, Radioactive Waste Page, http://www.nrc.gov/NRC/radwaste.

RADIATION AND RADIOACTIVE WASTES

International Atomic Energy Agency, "Managing Radioactive Waste" Fact Sheet, http://www.iaea.org/worldatom/inforesource/factsheets/manradwa.html.

National Research Council, Board on Radioactive Waste Management, http://www4.nas.edu/brwm/brwm-res.nsf.

U.S. Department of Energy, Office of Civilian Radioactive Waste Management, http://www. rw.doe.gov.

U.S. Environmental Protection Agency, Mixed-Waste Homepage, http://www.epa.gov/radiation/mixed-waste.

U.S. Nuclear Regulatory Commission, Radioactive Waste Page, http://www.nrc.gov/NRC/radwaste.

RADON

Radon in Earth, Air, and Water, http://sedwww.cr.usgs.gov:8080/radon/radonhome.html.

RAIN FORESTS

Greenpeace International, forests, http://www.greenpeace.org/~forests.

Rainforest Action Network (RAN)

President Randy Hayes
221 Pine Street Suite 500
San Francisco, CA 94104
Telephone: (415) 398-4404
Website: http://www.ran.org

Rainforest Alliance (RA)

65 Bleeker Street
New York, NY 10012
Website: http://www.rainforest-alliance.org

U.S. Forest Service, http://www.fs.fed.us.

World Wildlife Fund (Worldwide Fund for Nature), Forests for Life Campaign, http://www.panda.org/forests4life.

RESOURCE CONSERVATION AND RECOVERY ACT

Homepage, http://www.epa.gov/epaoswer/hotline.

SALMON

National Marine Fisheries Service, http://www.nwr.noaa.gov/1salmon/salmesa/index.htm.
NOAA Fisheries, http://www.nmfs.gov/.

SALT MARSHES

National Wetlands Research Center, http://www.nwrc.usgs.gov/educ_out.html.

USGS Coastal and Marine Geology, http://marine.usgs.gov/.

SANITARY LANDFILLS

Solid waste management, http://web.mit.edu/urbanupgrading/urban environment/

Landfills - Solid and Hazardous Waste and Groundwater Quality Protection, http://www.gfredlee.com/plandfil2.htm

SIBERIA

Siberia, http://www.cnit.nsk.su/univer/english/siberia.htm.

SOLAR ENERGY

American Solar Energy Society

2400 Central Avenue, Suite G-1
Boulder, CO 80301.
Website: http://www.soton.ac.uk/~solar/.

Solar Energy Industries Association

122 C Street, NW, 4th Floor
Washington, D.C. 20001.
Website: http://www.seia.org/main.htm.

U.S. Department of Energy, Photovoltaic Program, http://www.eren.doe.gov/pv/text_frameset.html.

SOLAR POND

Department of Mechanical and Industrial Engineering

University of Texas at El Paso
El Paso, TX 79968.
E-mail: aswift@cs.utep.edu.

SPENT FUEL

Environmental Protection Agency, www.ntp.doe.gov, www.rw.doe.gov/pages/resource/facts/transfct.htm.

SUPERFUND

Environmental Protection Agency, http://www.epa.gov/epaoswer/hotline.

Recycled Superfund sites, http://www.epa.gov/superfund/programs/recycle/index.htm.

Superfund Information, http://www.epa.gov/superfund.

U.S. EPA Superfund Program Homepage, Website: http://www.epa.gov/superfund/index.htm.

TENNESSEE VALLEY AUTHORITY

Homepage, http://www.tva.gov.

THOREAU, HENRY

Website: http://www.walden.org.

TOXIC CHEMICALS

Environmental Defense Fund, http://www.scorecard.org.

U.S. Department of Health and Human Services, Agency for Toxic Substances and Disease Registry (ASTDR), http://www.atsdr.cdc.gov/

U.S. Environmental Protection Agency, Integrated Risk Information System (IRIS), http://www.siri.org/msds; http://www.ilpi.com/mads/index.html.

U.S. Occupation Health and Safety Administration, http://www.toxicsubstances/index.html.

TOXIC RELEASE INVENTORY

Environmental Defense Fund, http://www.scorecard.org.
Environmental Protection Agency, http://www.epa.gov.

Teach with Databases, Toxic Release Inventory, http://www.nsta.org/pubs/special/pb143x01.htm.

TOXIC WASTE

Environmental Defense Fund, http://www.scorecard.org.

Institute for Global Communications, http://www.igc.org/igc/issues/tw/.

TRADE RECORDS ANALYSIS OF FLORA AND FAUNA IN COMMERCE (TRAFFIC)

Homepage, http://www.traffic.org/about/.

URBAN FORESTS

American Forests, http://www.amfor.org.

TreeLink, http://www.treelink.org.

VERTEBRATES

U.S. Fish and Wildlife Service, Species List of Endangered and Threatened Wildlife, http://www.fws.gov/r9endspp/lsppinfo.html.

VICUNA

U.S. Fish and Wildlife Service, Species List of Endangered and Threatened Wildlife, http://endangered.fws.gov

VITRIFICATION

U.S. Department of Energy, http://www.em.doe.gov/fs/fs3m.html.

VOLCANOES

USGS, Volcanoes in the Learning Web, http://www.usgs.gov/education/learnweb/volcano/index.html.

Volcano Hazards, http://volcanoes.usgs.gov/.

WATER CONSERVATION AND POLLUTION

Early History of the Clean Water Act, http://epa.gov/history/topics.

Environmental Protection Agency, Office of Wetlands, Oceans, Watersheds for Nonpoint Source information, http://www.epa.gov/owow/wetlands/wetland2.html; http://www.epa.gov/swerosps/ej/.

U.S. Geological Survey, Water Resources of the United States, National Groundwater Association Homepage, http://www.h2o-ngwa.org.

Water Resources Information, http://water.usgs.gov/.

Water Use Data, http://water.usgs.gov/public/watuse/.

WETLANDS

National Wetlands Research Center, http://www.nwrc.usgs.gov/educ_out.html.

Ramsar Convention on Wetlands (International), http://www2.iucn.org/themes/ramsar/.

Ramsar List of Wetlands of International Importance, http://ramsar.org/key_sitelist.htm.

WHALES

Institute of Cetacean Research (ICR), http://www.whalesci.org.

U.S. Fish and Wildlife Service, Species List of Endangered and Threatened Wildlife, http://www.fws.gov/r9endspp/lsppinfo.html; http://www.highnorth.no/iceland/th-in-to.htm; http://greenpeace.org/.

WILDERNESS

U.S. Forest Service, *Roadless Area Review and Evaluation*, http://www.fs.fed.us.

Wilderness Society, http://www.wilderness.org/newsroom/factsheets.htm.

WILDLIFE REFUGES

Conservation International, http://www.conservation.org.

Nature Conservancy, http://www.tnc.org.

U.S. Fish and Wildlife Service, National Wildlife Refuge System, http://refuges.fws.gov.

World Conservation Union/International Union for the Conservation of Nature, http://www.iucn.org.

WIND ENERGY

American Wind Energy Association

122 C Street NW, 4th Floor
Washington, D.C. 20001
Telephone: (202) 383-2500.
E-mail: awea@mcimail.com.
Website: http://www.awea.org.

Center for Renewable Energy and Sustainable Technology (CREST)

Solar Energy Research and Education Foundation
777 North Capitol Street NE, Suite 805
Washington, D.C. 20002
Website: http://solstice.crest.org/.

WOLVES

U.S. Fish and Wildlife Service, http://www.fws.gov/.
U.S. Fish and Wildlife Service, Species List of Endangered and Threatened Wildlife, http://endangered.fws.gov/.

World Wildlife Fund

1250 24th Street, NW
Washington, D.C. 20037
Telephone: 1-800-225-5993
Website: http://www.worldwildlife.org/.

WORLD HEALTH ORGANIZATION

Homepage, http://www.who.int.

WORLD WILDLIFE FUND

1250 24th Street, NW
Washington, D.C. 20037
Telephone: 1-800-225-5993
Website: http://www.wwf.org/.

YUCCA MOUNTAIN PROJECT

Homepage, http://www.ymp.gov/.

ZEBRAS

U.S. Fish and Wildlife Service, Species List of Endangered and Threatened Wildlife, http://endangered.fws.gov/.

ZOOS

Bronx Zoo, http://www.bronxzoo.com/.
San Diego Zoo, http://www.sandiegozoo.org/.

Appendix D: Environmental Organizations

Action for Animals

P.O. Box 17702
Austin, TX 78760
Telephone: (512) 416-1617
Fax: (512) 445-3454
Website: http://www.envirolink.org/

African Wildlife Foundation (AWF)

1400 Sixteenth Street, NW, Suite 120
Washington, D.C. 20036
Telephone: (202) 939-3333
Fax: (202) 939-3332
Website: http://www.awf.org/home.html

Agency for Toxic Substances and Diseases, Registry Division of Toxicology (ATSDR)

1600 Clifton Road
NE Mailstop E-29
Atlanta, GA 30333
Telephone: (888) 42-ATSDR or (888) 422-8737
E-mail: ATSDRIC@cdc.gov
Website: http://www.atsdr.cdc.gov/contacts.html

Alaska Forum for Environmental Responsibility

P.O. Box 188
Valdez, AK 99686
Telephone: (907) 835-5460
Fax: (907) 835-5410
Website: http://www.accessone.com/~afersea

American Conifer Society (ACS)

P.O. Box 360
Keswick, VA 22947-0360
Telephone: (804) 984-3660
Fax: (804) 984-3660
E-mail: ACSconifer@aol.com
Website: http://www.pacificrim.net/~bydesign/acs.html

American Forests

P.O. Box 2000
Washington, D.C. 20013
Telephone: (202) 955-4500
Website: http://www.americanforests.org

American Nuclear Society

555 North Kensington Avenue
La Grange Park, IL 60525
Telephone: (708) 352-6611
Fax: (708) 352-0499
E-mail: NUCLEUS@ans.org
Website: http://www.ans.org

American Oceans Campaign

201 Massachusetts Avenue NE, Suite C-3
Washington, D.C. 20002
Telephone: (202) 544-3526
Fax: (202) 544-5625
E-mail: aocdc@wizard.net
Website: http://www.americanoceans.org

American Rivers

1025 Vermont Avenue NW, Suite 720
Washington, D.C. 20005
Telephone: (202) 347-7500
Fax: (202) 347-9240
E-mail: amrivers@amrivers.org
Website: http://www.amrivers.org

American Society for Horticultural Science (ASHS)

600 Cameron Street
Alexandria, VA 22314-2562

Telephone: (703) 836-4606
Fax: (703) 836-2024
E-mail: webmaster@ashs.org
Website: http://www.ashs.org

American Society for the Prevention of Cruelty to Animals (ASPCA)

424 East Ninety-second Street
New York, NY 10128
Telephone: (212) 876-7700
Website: http://www.aspca.org

American Solar Energy Society

2400 Central Avenue, Suite G-1
Boulder, CO 80301
Telephone: (303) 443-3130
Fax: (303) 443-3212
E-mail: ases@ases.org
Website: http://www.ases.org
Publication: *Solar Today*

American Wind Energy Association

122 C Street NW, Fourth Floor
Washington, D.C. 20001
Telephone: (202) 383-2500
E-mail: awea@mcimail.com
Website: http://www.awea.org

Animal Legal Defense Fund (ALDF)

127 Fourth Street
Petaluma, CA 94952
Telephone: (707) 769-7771
Fax: (707) 769-0785
E-mail: info@aldf.org
Website: http://www.aldf.org

Animal Rights Network

P.O. Box 25881
Baltimore, MD 21224
Telephone: (410) 675-4566
Fax: (410) 675-0066
Website: http://www.envirolink.org/arrs/aa/index.html
Publication: *Animals' AGENDA*, a bimonthly magazine

Baron's Haven Freehold

104 South Main Street
Cadiz, OH 43907
Telephone: (740) 942-8405
Website: http://bhfi.1st.net

Biodiversity Support Program (BSP)

1250 North Twenty-fourth Street NW, Suite 600
Washington, D.C. 20037
Telephone: (202) 778-9681
Fax: (202) 861-8324
Website: http://www.BSPonline.org

Biosfera

Pres. Vargas 435, Suites 1104 and 1105
Rio de Janeiro, RJ 20077-900
Brazil

Birds of Prey Foundation

2290 South 104th Street
Broomfield, CO 80020
Telephone: (303) 460-0674
Fax: (303) 666-1050
E-mail: raptor@birds-of-prey.org
Website: http://www.birds-of-prey.org

Build the Earth

3818 Surfwood Road
Malibu, CA 90265
Telephone: (310) 454-0963

Center for Conversion and Research of Endangered Wildlife (CREW)

Cincinnati Zoo and Botanical Garden
3400 Vine Street
Cincinnati, OH 45220
E-mail: terri.roth@cincyzoo.org

Center for Marine Conservation

1725 DeSales Street SW, Suite 600
Washington, D.C. 20036
Telephone: (202) 429-5609
Fax: (202) 872-0619
E-mail: cmc@dccmc.org
Website: http://www.cmc-ocean.org

Centers for Disease Control (CDC)

1600 Clifton Rd.
Atlanta, GA 30333

Telephone: (800) 311-3435
Website: http://www.cdc.gov

Cheetah Conservation Fund (CCF)

P.O. Box 1380
Ojai, CA 93024
Telephone: (805) 640-0390
Fax: (815) 640-0230
E-mail: info@cheetah.org
Website: http://www.cheetah.org

Clean Air Council (CAC)

135 South Nineteenth Street, Suite 300
Philadelphia, PA 19103
Telephone: (888) 567-7796
Website: http://www.libertynet.org/~cleanair/

Coalition for Economically Responsible Economies (CERES)

11 Arlington Street, Sixth Floor
Boston, MA 02116-3411
Telephone: (617) 247-0700
Fax: (617) 267-5400
Website: http://www.ceres.org

Conservation International

1015 Eighteenth Street NW Suite 1000
Washington, D.C. 20036
Telephone: (202) 429-5660
Website: http://www.conservation.org/
Publication: *Orion Nature Quarterly*

Convention on International Trade in Endangered Species of Wild Fauna and Flora (CITES)

CITES Secretariat
International Environment House, 15, chemin des Anémones, CH-1219
Châtelaine-Geneva, Switzerland
E-mail: cites@unep.ch
Website: http://www.cites.org/index.shtml

Council for Responsible Genetics

5 Upland Road, Suite 3
Cambridge, MA 02140
Website: http://www.gene-watch.org

Cousteau Society

870 Greenbriar Circle, Suite 402
Chesapeake, VA 23320
Telephone: (804) 523-9335
E-mail: cousteau@infi.net
Website: http://www.cousteausociety.org/
Publication: *Calypso Log*

Defenders of Wildlife

1101 Fourteenth Street NW, Room 1400
Washington, D.C. 20005
Telephone: (800) 441-4395
Website: http://www.Defenders.org
Publication: *Defenders*, a quarterly magazine

Dian Fossey Gorilla Fund International

800 Cherokee Avenue SE
Atlanta, GA 30315-1440
Telephone: (800) 851-0203
Fax: (404) 624-5999
E-mail: 2help@gorillafund.org
Website: http://www.gorillafund.org/000_core_frmset.html

Earth Day Network

1616 P Street NW
Suite 200
Washington, D.C. 20036
E-mail: earthday@earthday.net
Website: http://www.earthday.net

Earth Island Institute (EII)

300 Broadway, Suite 28
San Francisco, CA 94133
Telephone: (415) 788-3666
Fax: (415) 788-7324
Website: http://www.earthisland.org/abouteii/abouteii.html
Publication: *Earth Island Journal*, a quarterly magazine

Earth, Pulp, and Paper

P.O. Box 64
Leggett, CA 95585
Telephone: (707) 925-6494
E-mail: tree@tree.org
Website: http://www.tree.org/epp.htm

EarthFirst! (EF!)

P.O. Box 5176
Missoula, MT 59806
Website: http://www.webdirectory.com/General_Environmental_Interest/Earth_First_/

Earthwatch Institute

In United States and Canada
3 Clocktower Place, Suite 100
Box 75
Maynard, MA 01754
Telephone: (800) 776-0188 or (617) 926-8200
Fax: (617) 926-8532
In Europe
57 Woodstock Road
Oxford OX2 6HJ, United Kingdom
E-mail: info@uk.earthwatch.org
Website: http://www.earthwatch.org

EcoCorps

1585 A Folsom Avenue
San Francisco, CA 94103
Telephone: (415) 522-1680
Fax: (415) 626-1510
E-mail: eathvoice@ecocorps.org
Website: http://www.owplaza.com/eco

Ecotourism Society

P.O. Box 755
North Bennington, VT 05257
Telephone: (802) 447-2121
Fax: (802) 447-2122
E-mail: ecomail@ecotourism.org
Website: http://www.ecotoursim.org

E. F. Schumacher Society

140 Jug End Road
Great Barrington, MA 01230
Telephone: (413) 528-1737
E-mail: efssociety@aol.com
Website: http://members.aol.com/efssociety/index.html

Electric Vehicle Association of the Americas

701 Pennsylvania Avenue NW, Fourth Floor
Washington, D.C. 20004
Telephone: (202) 508-5995
Fax: (202) 508-5924
Website: http://www.evaa.org

Environmental Defense Fund (EDF)

257 Park Avenue South
New York, NY 10010
Telephone: (800) 684-3322
Fax: (212) 505-2375
E-mail (for general questions and information): Contact@environmentaldefense.org
Website: http://www.edf.org
Publication: *Nature Journal*, a monthly magazine

Exotic Cat Refuge and Wildlife Orphanage

Route 3, Box 96A
Kirbyville, TX 75956
Telephone: (409) 423-4847

Federal Emergency and Management Agency (FEMA)

500 C Street SW
Washington, D.C. 20472
Website: http://www.fema.gov

Friends of the Earth (FOE)

1025 Vermont Avenue NW, Suite 300
Washington, D.C. 20005-6303
Telephone: (202) 783-7400
Fax: (202) 783-0444
E-mail: foe@foe.org
Website: http://www.foe.org

Green Seal

1001 Connecticut Avenue NW, Suite 827
Washington, D.C. 20036-5525
Telephone: (202) 872-6400
Fax: (202) 872-4324
Website: http://www.greenseal.org

Greenpeace USA

1436 U Street NW
Washington, D.C. 20009
Telephone: (202) 462-1177
Website: http://www.greenpeaceusa.org/
Publication: *Greenpeace Magazine*

Hawkwatch International

P.O. Box 660
Salt Lake City, UT 84110
Telephone: (801) 524-8511
E-mail: hawkwatch@charitiesusa.com
Website: http://www.vpp.com/HawkWatch

Humane Society of the United States (HSUS)

2100 L Street NW
Washington, D.C. 20037
Website: http://www.hsus.org
Publications: *All Animals*, a quarterly magazine

International Atomic Energy Commission

P.O. Box 100
Wagramer Strasse 5
A-1400, Vienna, Austria
E-mail: Official.Mail@iaea.org
Website: http://www.iaea.org

International Council for Local Environmental Initiatives (ICLEI)

World Secretariat
16th Floor, West Tower, City Hall
Toronto, M5H 2N2, Canada
Fax: (416) 392-1478
Email: iclei@iclei.org
Website: http://www.iclei.org

International Rhino Foundation (IRF)

14000 International Road
Cumberland Ohio 43732
E-mail: IrhinoF@aol.com
Website: http://www.rhinos-irf.org

International Whaling Commission (IWC)

The Red House
135 Station Road
Impington, Cambridge CB4 9NP, United Kingdom
E-mail: iwc@iwcoffice.org
Website: http://ourworld.compuserve.com/homepages/iwcoffice

International Wolf Center

1396 Highway 169
Ely, MN 55731-8129
Telephone: (218) 365-4695
Fax: (218) 365-3318
Website: http://www.wolf.org

Jane Goodall Institute (JGI)

P.O. Box 14890
Silver Spring, MD 20911-4890
Telephone: (301) 565-0086
Fax: (301) 565-3188
E-mail: JGIinformation@janegoodall.org

Keep America Beautiful

1010 Washington Boulevard
Stamford, CT 06901
Telephone: (203) 323-8987
Fax: (203) 325-9199
E-mail: info@kab.org

League of Conservation Voters

1707 L Street, NW, Suite 750
Washington, D.C. 20036
Telephone: (202) 785-8683
Fax: (202) 835-0491
E-mail: lcv@lcv.org
Website: http://www.lcv.org

Mountain Lion Foundation (MLF)

P.O. Box 1896
Sacramento, CA 95812
Telephone: (916) 442-2666
E-mail: MLF@moutainlion.org
Website: http://www.mountainlion.org

National Alliance of River, Sound, and Bay Keepers

P.O. Box 130
Garrison, NY 10524
Telephone: (800) 217-4837
E-mail: keepers@keeper.org
Website: http://www.keeper.org

National Anti-Vivisection Society (NAVS)

53 West Jackson Street, Suite 1552
Chicago, IL 60604
Telephone: (800) 888-NAVS
E-mail: navs@navs.org
Website: http://www.navs.org

National Arbor Day Foundation

100 Arbor Avenue
Nebraska City, NE 68410
Telephone: (402) 474-5655
Website: http://www.arborday.org
Publication: *Arbor Day*, a bimonthly magazine

National Audubon Society (NAS)

700 Broadway
New York, NY 10003
Telephone: (212) 979-3000
Website: http://www.audubon.org
Publication: *Audubon*, a bimonthly magazine

National Center for Environmental Health

Mail Stop F-29
4770 Buford Highway NE
Atlanta, GA 30341-3724
Telephone NCEH Health Line: (888) 232-6789
Website: http://www.cdc.gov/nceh/ncehhome.htm

National Parks and Conservation Association (NPCA)

1015 Thirty-first Street NW
Washington, D.C. 20007
Telephone: (202) 944-8530; (800) NAT-PARK
E-mail: npca@npca.org
Website: http://www.npca.org
Publication: *National Parks*, a bimonthly magazine

National Wildlife Federation (NWF)

8925 Leesburg Pike
Vienna, VA 22184-0001
Telephone: (800) 822-9919
Website: http://www.nwf.org
Publication: *National Wildlife*, a bimonthly magazine

Natural Resources Defense Council (NRDC)

40 West Twentieth Street
New York, NY 10011
Website: http://www.nrdc.org
Publications: *Amiscus Journal*, a quarterly magazine

Nature Conservancy (TNC)

1815 North Lynn Street
Arlington, VA 22209
Telephone: (703) 841-5300
Fax: (703) 841-1283
Website: http://www.tnc.org
Publication: *Nature Conservancy*, a magazine

Noise Pollution Clearinghouse

P.O. Box 1137
Montpelier, VT 05601-1137
Telephone: (888) 200-8332
Website: http://www.nonoise.org

North Sea Commission

Business and Development Office
Skottenborg 26, DK-8800 Viborg, Denmark
Website: http:\\www.northsea.org

People for Animal Rights

P.O. Box 8707
Kansas City, MO 64114
Telephone: (816) 767-1199
E-mail: parinfo@envirolink.org
Website: http://www.parkc.org

People for the Ethical Treatment of Animals (PETA)

501 Front Street
Norfolk, VA 23510
Telephone: (757) 622-PETA
Fax: (757) 622-0457
Website: http://www.peta-online.org/

Orangutan Foundation International

822 South Wellesley Avenue
Los Angeles, CA 90049
Telephone: (800) ORANGUTAN
Fax: (310) 207-1556
E-mail: ofi@orangutan.org
Website: http://www.ns.net/orangutan

Ozone Action

1700 Connecticut Avenue NW, Third Floor
Washington, D.C. 20009
Telephone: (202) 265-6738

E-mail: cantando@essential.org
Website: www.ozone.org

Peregrine Fund

566 West Flying Hawk Lane
Boise, ID 83709
Telephone: (208) 362-3716
Fax: (208) 362-2376
E-mail: tpf@peregrinefund.org
Website: http://www.peregrinefund.org

Rachel Carson Council

8940 Jones Mill Road
Chevy Chase, MD 20815
Telephone: (301) 652-1877
E-mail: rccouncil@aol.com
Website: http://members.aol.com/rccouncil/ourpage

Rainforest Action Network

221 Pine Street, Suite 500
San Francisco, CA 94104-2740
Telephone: (415) 398-4404
Fax: (415) 398-2732
E-mail: rainforest@ran.org
Website: http://www.ran.org

Range Watch

45661 Poso Park Drive
Posey, CA 93260
Telephone: (805) 536-8668
E-mail: rangewatch@aol.com
Website: http://www.rangewatch.org

Raptor Resource Project

2580 310th Street
Ridgeway, IA 52165
E-mail: rrp@salamander.com
Website: http://www.salamander.com~rpp

Reef Relief

201 William Street
Key West, FL 33041
Telephone: (305) 294-3100
Fax: (305) 923-9515
E-mail: reef@bellsouth.net
Website: http://www.reefrelief.org

ReefKeeper International

2809 Bird Avenue, Suite 162
Miami, FL 33133
Telephone: (305) 358-4600
Fax: (305) 358-3030
E-mail: reefkeeper@reefkeeper.org
Website: http://www.reefkeeper.org

Renewable Energy Policy Project-Center for Renewable Energy and Sustainable Technology (REPP-CREST)

National Headquarters
1612 K Street, NW, Suite 202
Washington, D.C. 20006
Website: http://www.solstice.crest.org

Resources for the Future (RFF)

1616 P Street NW
Washington, D.C. 20036
Telephone: (202) 328-5000
Fax: (202) 939-3460
E-mail: info@rff.org
Website: http://www.rff.org

Roger Tory Peterson Institute

311 Curtis Street
Jamestown, NY 14701
Telephone: (716) 665-2473
E-mail: webmaster@rtpi.org

Sierra Club

85 Second Street, Second Floor
San Francisco, CA 94105
Telephone: (415) 977-5630
Fax: (415) 977-5799
E-mail (general information): information@sierraclub.org
Website: http://www.Sierraclub.org
Publication: *Sierra*, a bimonthly magazine

Smithsonian Institution Conservation & Research Center (CRC)

Website: http://www.si.edu/crc/brochure/index.htm

Society of American Foresters

5400 Grosvenor Lane
Bethesda, MD 20814

Telephone: (301) 897-8720
Fax: (301) 897-3690
E-mail: safweb@safnet.org
Website: http://www.safnet.org

Surfrider Foundation USA

122 South El Camino Real, Suite 67
San Clemente, CA 92672
Telephone: (949) 492-8170
Fax: (949) 492-8142
Website: http://www.surfrider.org

Union of Concerned Scientists

National Headquarters
2 Brattle Square
Cambridge, MA 02238
Telephone: (617) 547-5552
E-mail: ucs@ucsusa.org
Website: http://www.ucsusa.org
Publications: *Nucleus*, a quarterly magazine; *Earthwise*, a quarterly newsletter

United Nations Environment Programme (Regional)

2 United Nations Plaza
NY, NY 10017
Telephone: (212) 963-8138
Website: http://www.unep.org

United Nations Food and Agriculture Organization (FAO)

Website: http://www.fao.org
Liaison office with North America
Suite 300, 2175 K Street NW, Washington D.C. 20437-0001

United Nations Man and the Biosphere Programme (UNMAB)

U.S. MAB Secretariat, OES/ETC/MAB
Department of State
Washington, D.C. 20522-4401
Website: http://www.mabnet.org

U.S. Department of Agriculture (USDA)

14th Street and Independence Avenue., SW,
Washington, D.C. 20250
Website: http://www.usda.gov

U.S. Department of Energy (DOE)

Forrestal Building
1000 Independence Avenue, SW,
Washington, D.C. 20585
Website: http://www.doe.gov

U.S. Environmental Protection Agency (EPA)

401 M Street SW
Washington, D.C. 20460
Website: http://www.epa.gov

U.S. Fish and Wildlife Service (FWS)

1849 C Street NW
Washington, D.C. 20240
Telephone: (202) 208-5634
Website: http://www.fws.org

U.S. Geological Survey (USGS)

U.S. Dept. of Interior
1849 C Street, NW
Washington, D.C. 20240
Website: http://www.usgs.gov

U.S. National Park Service (NPS)

U.S. Dept. of Interior
1849 C Street, NW
Washington, D.C. 20240
Website: http://www.nps.gov

U.S. Nuclear Regulatory Commission (NRC)

One White Flint North
11555 Rockville Pike
Rockville, Maryland 20852
Website: http://www.nrc.gov

Wilderness Society

900 Seventeenth Street NW
Washington, D.C. 20006-2506
Telephone: (800) THE-WILD
Website: www.wilderness.org

Wildlands Project (TWP)

1955 West Grant Road, Suite 145
Tucson, AZ 85745
Telephone: (520) 884-0875
Fax: (520) 884-0962

E-mail: information@twp.org
Website: http://www.twp.org

World Conservation Monitoring Centre (WCMC)

219 Huntington Road
Cambridge CB3 ODL, United Kingdom
E-mail: info@wcmc.org.uk
Website: http://www.wcmc.org.uk

World Conservation Union (IUCN)

1630 Connecticut Avenue NW, Third Floor
Washington, D.C. 20009-1053
Telephone: (202) 387-4826
Fax: (202) 387-4823
E-mail: postmaster@iucnus.org
Website: http://www.iucn.org

World Health Organization (WHO)

Avenue Appia 20
1211 Geneva 27
Switzerland
Website: http://www.eho.int
E-mail: inf@who.int

World Parrot Trust United States

P.O. Box 50733
Saint Paul, MN 55150
Telephone: (651) 994-2581
Fax: (651) 994-2580
E-mail: usa@worldparrottrust.org

United Kingdom

Karen Allmann, Administrator,
Glanmor HouseHayle,
Cornwall TR27 4HY,
United Kingdom
E-mail: uk@worldparrottrust.org

Australia

Mike Owen
7 Monteray Street
Mooloolaba, Queensland 4557, Australia
E-mail: australia@worldparrottrust.org
Website: http://www.world parrottrust.org

World Resources Institute

1709 New York Avenue NW
Washington, D.C. 20006
Telephone: (202) 638-6300
E-mail: info@wri.org
Website: http://www.wri.org/wri/biodiv

World Society for the Protection of Animals (WSPA)

P.O. Box 190
Jamaica Plain, MA 02130
Website: http://www.wspa.org

United Kingdom Division
Website: http://www.wspa.org.uk/home.html

World Wildlife Fund, US (WWF)

1250 Twenty-fourth Street NW
P.O. Box 97180
Washington, D.C. 20077-7180
Telephone: (800) CALL-WWF
Website: http://www.worldwildlife.org

WorldWatch Institute

1776 Massachusetts Avenue NW
Washington, D.C. 20036
Telephone: (202) 452-1999
Website: http://www.worldwatch.org/
Publications: *WorldWatch, State of the World, Vital Signs* (annuals)

Zero Population Growth

1400 Sixteenth Street NW, Suite 320
Washington, D.C. 20036
Telephone: (202) 332-2200
Fax: (202) 332-2302
E-mail: zpg@igc.apc.org
Website: http://www.zpg.org

Zoe Foundation

983 River Road
Johns Island, SC 29455
Telephone: (803) 559-4790
E-mail: savage@awod.com
Website: http://www.2zoe.com

INDEX

f indicates figures and photos; t indicates tables

1: Earth Systems, Volume I
2: Resources, Volume II
3: People, Volume III
4: Human Impact, Volume IV
5: Sustainable Society, Volume V

1: Earth Systems, Volume I
2: Resources, Volume II
3: People, Volume III
4: Human Impact, Volume IV
5: Sustainable Society, Volume V

1: Earth Systems, Volume I **2:** Resources, Volume II **3:** People, Volume III **4:** Human Impact, Volume IV **5:** Sustainable Society, Volume V

1: Earth Systems, Volume I
2: Resources, Volume II
3: People, Volume III
4: Human Impact, Volume IV
5: Sustainable Society, Volume V

1: Earth Systems, Volume I **2:** Resources, Volume II **3:** People, Volume III **4:** Human Impact, Volume IV **5:** Sustainable Society, Volume V

1: Earth Systems, Volume I **2:** Resources, Volume II **3:** People, Volume III **4:** Human Impact, Volume IV **5:** Sustainable Society, Volume V

ABOUT THE AUTHORS

JOHN MONGILLO is a noted science writer and educator. He is coauthor of *Encyclopedia of Environmental Science*, and *Environmental Activists*, both available from Greenwood.

PETER MONGILLO has won several awards for his teaching, including School District Teacher of the Year, National Endowment for the Humanities Fellowship Award, and the National Council for Geographic Education Distinguished Teacher Award.